Rapid Prototyping
and
Tooling Research

Collected papers from the *First National Conference on Rapid Prototyping and Tooling Research*, held at Buckinghamshire College, UK, 6th–7th November 1995.

The conference initiated by the Rapid Prototyping Forum, was organized in association with the Institution of Mechanical Engineers.

Editor

Graham Bennett
Department of Engineering Technology
Buckinghamshire College of Higher Education
UK

Editorial Board

First National Conference on
Rapid Prototyping
and
Tooling Research

Edited by

Graham Bennett
Department of Engineering Technology
Buckinghamshire College, UK

Mechanical Engineering Publications Limited
London and Bury St Edmunds, UK

First published 1995

ISBN 0 85298 982 2

© The Rapid Prototyping Forum

A CIP catalogue record for this book is available from the British Library.

Printed by Antony Rowe Limited, Chippenham Wiltshire, UK

Foreword

'Rapid Prototyping and Tooling' is an oft quoted but ill defined technology. For the purposes of this publication it shall mean those activities and processes which significantly reduce the time between concept and the production of high quality prototype, parts or tools. In particular, we shall primarily be concerned with methods of producing hardware, either 'real' parts or working tools via techniques of layer manufacture. Nevertheless, some work is included which reports methods of improving the efficiency of the purely CAD aspects of the product design and development process.

It is probably true that competition between companies involved in the manufacture of goods, particularly consumer products has never been more fierce.

The rapidity of technological innovation and development is such that it is unwise for a company not to maintain a knowledge of the most up-to date development and manufacturing practices. A company which quickly adopts new and effective practices can receive a very significant advantage over its competitors.

Rapid Prototyping and Tooling (RP&T) is one area where significant competitive benefits are gained by those companies which adopt it.

Furthermore, for the industry of a country to maintain its competitive position in the world, it is essential that a foundation of expertise in these technologies exist within its research and development community. It is therefore reassuring to know that there is an active, if at the present time somewhat small, research and development community in Rapid Prototyping and Tooling technologies within the UK. The work undertaken in the UK is, in the majority based upon secondary processes, i.e. in the production of tools and parts from master 'formers' made by existing machines. It would be a welcome development for work to be undertaken in the means by which these initial 'formers' can be made.

The Conference from which these proceedings have been taken arose as a result of conversations between a group of UK Universities and industrial RP&T users. It is intended that the Conference will become an annual event, providing a forum for UK workers, particularly young researchers, to present their work within the UK and thereby assisting the promotion of the understanding, use and implementation of this technology within UK research institutions and companies.

I would like to extend my very sincere thanks to the editorial board, without whose excellent help, the refereeing of these works would not have been possible. I would also like to acknowledge the long hours and hard work put into this project by Mrs Sharon Hoggarth.

Finally, I would like to add the comment that we are at the beginning of what is likely to be a revolution of the manner by which products are designed and manufactured. What are simply quick methods of producing prototypes and relatively low production volume tools will undoubtedly be developed into methods of large scale production. Techniques which will have surprisingly low lead times for such production are but a few years away.

The ramifications of this are difficult to predict, but I believe that a good deal of manufacturing processes twenty years from now will be transformed, in terms of speed and cost. It is essential, if companies are to remain competitive, that they educate themselves in these technologies, support their development, and are willing to adopt them sooner rather than later. The prize of continuing profitability will only go to those who do.

Graham R Bennett
Centre for Rapid Design and Manufacture
Buckinghamshire College
High Wycombe
UK

September, 1995

Contents

An investigation into the feasibility of rapid prototyping technologies being adopted during industrial design activity

G CHESHIRE
Staffordshire University, UK
A EVANS and **P W WORMALD**
Loughborough University, UK

SYNOPSIS

The practice of industrial design involves the integration of engineering elements of new product development with a visual form and user interface appropriate to a particular market. Whilst having a general engineering capability, the specialism of the industrial designer is predominantly visual and ergonomic.

Until relatively recently there was little impetus for the industrial designer to adopt computer aided design techniques as the systems offered inadequate advantages over 'conventional' practices to justify the financial investment. However, the advent of rapid prototyping is set to bring about a serious rethink on the role of computer modelling by industrial designers.

This paper investigates the changes in working practices necessary to exploit the rapid prototyping technologies currently available, and identifies the benefits brought about by this change. The 'conventional' industrial design practices are examined, and contrasted with a methodology required to combine elements of this with the generation of computer solid models necessary for rapid prototyping.

The paper identifies the changes in methodology necessary to enable industrial designers to exploit rapid prototyping, and the benefits that this will generate.

The findings are derived from a research exercise involving the industrial design of an innovatory nylon line garden trimmer for domestic use.

1 INTRODUCTION

The techniques of presentation, unique to the industrial designer, rely on 2D visualisation and 3D models, the nature of which has remained largely unchanged since the beginnings of the profession at the start of this century.

The impact of Computer Aided Design (CAD) has had relatively little impact on the working practices of industrial designers. Where used, the applications have been generally to aid drafting. At the core of industrial design activity is a requirement to produce physical models, which until relatively recently could only be undertaken by model makers interpreting engineering drawings. With the introduction of rapid prototyping, the viability of producing physical models from computer models represents a significant challenge to the 'conventional' working practices of the profession.

The findings are derived primarily from design work carried out on a family of domestic garden products, notably an electric lawn mower and a nylon line garden trimmer. The mower was designed using 'conventional' industrial design methodologies. To fully appreciate the benefits that rapid prototyping could make to the industrial design profession, it is first necessary to define what is regarded as a 'conventional' industrial design strategy for new product development using 2D and 3D modelling techniques. Therefore the mower design process is used to illustrate this.

2 'CONVENTIONAL' INDUSTRIAL DESIGN METHODOLOGY .

Central to industrial design creative work is modelling. Modelling is that activity whereby products are represented in a form other than in their final manufactured state. All design disciplines use models. One of the most common forms of model is the engineering drawing in which all aspects of a product (form, finishes, construction details etc.) can be represented in 2D. A commonly understood form of model is the 3D physical model. This could be a non-functioning appearance model, a functional model to develop some technological feature, or a pre-production prototype which is very close in appearance, construction and function to the manufactured item. The different disciplines involved in the process of new product development (NPD) have their own forms of models to aid the progression of their tasks. These models are critical to the success of NPD.

To fully appreciate the benefits that rapid prototyping could make to the industrial design profession, it is first necessary to define what is regarded as a 'conventional' industrial design methodology for new product development using 2D and 3D modelling techniques. The following is a list of model types and a description of their role in industrial design activity.

2.1 Sketch (2D)

The ability to sketch is a fundamental skill for the industrial designer. It is the process whereby three dimensional relationships can be quickly and effectively manipulated using a two dimensional medium. Using little more than pencil and paper it is possible via sketching to externalise, evaluate, and develop a concept until it reaches a level of acceptability (figure 1).

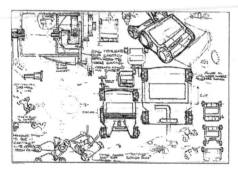

Fig 1 Mower sketch drawing   Fig 2 Mower sketch model in foam

2.2 Sketch Model (3D)

Whilst undertaking design activity it is often necessary to translate drawings into a more representative 3D format. Materials such as card and foam (figure 2) can be quickly manipulated to give approximations of form that enable the designer to understand more fully the complex relationships between components, cavities, interfaces and form.

This first stage of 3D physical modelling may take place at any time in the design development process but often occurs during the early stages.

2.3   Development Drawing (2D)

There is rarely a clear-cut demarcation between a Sketch and Development Drawing, but at some stage in the design process there is a need to establish greater control and accuracy of representation. Instead of continuing with a predominantly freehand, monochromatic technique as typified by Sketching, Development Drawings make use of drawing equipment and colour (figure 3).

The added realism afforded by Development Drawings facilitates an enhanced appreciation of product attributes, and continue to be employed until more formal 'Renderings' are produced.

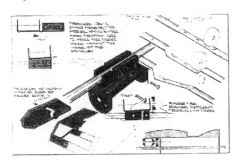

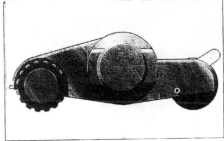

Fig 3 Mower development drawing   Fig 4 Mower rendering

2.4   Rendering (2D)

Having generated ideas using highly interactive 2D Sketching and 3D Sketch Modelling, and refined these via Development Drawings, a more formal Rendering is produced to communicate the design to a client or senior management team.

Renderings depict the major features of a product proposal as accurately as possible (figure 4). This requires considerable skill in the generation of perspective drawings and application of colour. As many designers operate to tight timescales, the media and techniques developed for the production of renderings facilitate rapid yet effective results.

2.5   Appearance Model (3D)

Appearance models are produced towards the end of the industrial design phase of NPD and are exact visual representations of the proposed product. Such models have accurate surface treatment in gloss or matt along with appropriate badging. Materials used include plastics (sheet, extruded sections), wood (Jelutong, Medium Density Fibreboard), and cellulose paints. Metals are generally used for the more structural components. The result is a visual impression of the final proposal, accurate in form and finish but not materials usage (figure 5). As Appearance Models are essentially hand made, they are notoriously expensive.

Appearance models rarely have any functional components as only the essential features necessary for an ergonomic and visual assessment are possible. This is due partly to the difficulty of creating moving surfaces with sufficient structural integrity to perform as the production item.

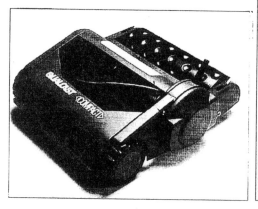

Fig 5 Mower appearance model

Fig 6 Mower appearance prototype

2.6 Appearance Prototype (3D)

The integration of the industrial design solution into a working prototype amounts to the highest level of 3D modelling in the NPD process. Fit and function can be assessed, along with accurate ergonomic evaluation (figure 6).

Unlike the Appearance Model, the Appearance Prototype requires representative wall thicknesses to create the internal cavities for components. The production of these simulated components (typically representing injection mouldings) necessitates their fabrication from a combination of vacuum formings and bonded sheet material.

The Appearance Prototype is also necessary to confirm the accuracy of the engineering drawings that are required to produce the production tooling. With the transfer of information through so many media there is always a possibility of human error affecting the integrity of the processes. Unfortunately such errors tend not to appear until the tooling has been commissioned, at which stage modifications may be difficult and expensive or even impossible.

3   A COMPUTER-BASED STRATEGY

The 'conventional' industrial design modelling procedures as described above are concerned with pursuing visual and ergonomic design objectives for the product. Engineering development tends to use different forms of product models to achieve its aims. These will include mock-ups of electrical or electronic systems, engineering layout drawings, and mechanical mock-ups.

Engineering and industrial design is carried out as part of the early stages of NPD and the shortest lead times are often achieved when the various activities incorporate as much concurrency as possible. The working relationships of the various designers and engineers is also a critical factor in the success of product development. The models used for product development play a large part in this relationship.

As a relatively recent approach to design modelling, rapid prototyping may significantly enhance the degree of concurrency that can be employed by industrial designers. All rapid prototyping technologies rely on 3D computer models to generate the physical models which have the capability for use in further engineering development and production tooling. The success of rapid prototyping in NPD therefore relies on the adoption of 3D computer modelling as a standard form of modelling in the design phase. This means that industrial designers must adopt CAD systems as primary modelling tools

Industrial designers have used simple computer aided industrial design (CAID) applications for product modelling for nearly a decade. This has typically been for 2D visualisation, but the computer models used for 2D visualisation do not normally have the geometrical accuracy or integrity for 3D engineering analysis or the capability for translation into computer numerical controlled (CNC) or rapid prototyping file formats. There are only a few 3D modelling applications created for industrial design activity that can generate the quality of 3D computer model necessary for rapid prototyping. These include applications such as DeskArtes and Alias Studio. The alternative to these are high-level integrated 3D computer aided engineering applications such as Pro-Engineer, SDRC I-DEAS, Unigraphics and IBM CATIA.

If industrial designers are to make use of rapid prototyping then they must be generating 3D computer models as part of their design methodology. At present in the UK, it is the exception rather than rule to find an industrial designer generating 3D computer models as a natural part of the design process. There are only a few isolated consultancies that have begun to specialise in producing computer models for CNC machining, rapid prototyping and direct model exchange with clients and mould tool makers.

4 CAID CASE STUDY

A methodology has been devised that allows the industrial designer to effectively integrate computer applications into professional practice and make effective use of rapid prototyping. This was devised from the findings of several design commissions undertaken by the authors. An earlier project was the design and production of a piece of ceramic tableware (Cheshire et al). CAID and rapid prototyping were used very successfully in the design and tooling production of the coffee pot. The authors gained valuable experience in the use of CAID but particularly in the usefulness of the stereolithography model as an industrial design model. That is as an equivalent to an appearance model but with physical qualities of weight and balance that added extra value beyond the conventional industrial design appearance model. The authors have added to the experience of the coffee pot with the study described fully in this paper, the design of a nylon line garden trimmer.

As both the nature of the products and management of the projects differed, it was possible to establish related and distinct findings that contributed to devising the overall methodology. The CAID application used throughout was DeskArtes running on a Silicon Graphics Unix workstation.

4.1 Nylon Line Garden Trimmer
Despite the application of high level industrial design software, it was impossible to replace the spontaneity afforded by 'conventional' Sketching at the beginning of the line trimmer project (figure 7). It was only when the basic elements of the product had been defined that CAID became appropriate.

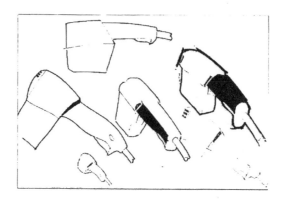

Fig 7 Trimmer sketch drawings

Once employed, the transition from the 'conventional' sketches to an extremely realistic CAID rendering was rapid (figure 8). This enabled relatively subtle design decisions to be made with an immediate and extremely accurate indication of their implications. This level of sophistication is not possible using 'conventional' techniques, where unexpected surface effects only become apparent when models are produced.

Fig 8 CAID rendering of nylon line garden trimmer

As design development continued it became apparent that the CAID system was starting to merge some of the 'conventional' Sketch, Development Drawing, and Rendering activities. This was all with the knowledge that engineering drawings would not be required at a later stage due to the geometrical integrity of the CAID surfaces. This developed to such a degree that the resulting forms arose out of the creative opportunities brought about by the CAID system, and a knowledge that the highly complex surfaces would not be physically modelled manually (as rapid prototyping was to be used). Figures 9 and 10 show details of the handle and cutter.

 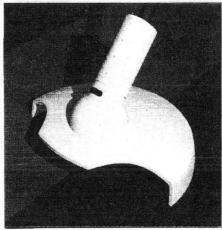

Fig 9 CAID rendering of trimmer hand grip   Fig 10 CAID rendering of trimmer cutter head

A key feature of the line trimmer was a necessity to comply with a detailed ergonomic specification. This resulted in a necessity to produce simple 3D 'Sketch Models' very early on in the design process so that the overall dimensions and configuration could be confirmed prior to computer modelling (figure 11). Whilst computer rendering may be extremely realistic there is no substitute for actually holding a physical model to assess ergonomics. As the weight balance of this product was an important ergonomic consideration, it would not be possible to accurately assess this until a 'conventional' Appearance Prototype had been produced. As the programme continues the authors will be replacing the 'conventional' Appearance Model and Appearance Prototype with a fully working rapid prototype.

5   CAID MODELLING METHODOLOGY

The case study research has identified that the processes carried out using computer based techniques and rapid prototyping are significantly related to the 'conventional' processes already described. Whilst not rejecting the most useful aspect of the 'conventional' route, changes are inevitable, and the following methodology draws on the case study findings.

5.1   Sketch

It was discovered that no computer based operations could reproduce the spontaneity afforded by 'conventional' paper and pencil techniques (figure 7). Sketching therefore formed the starting point for computer modelling as it does for 'conventional' practices.

5.2   Computer Sketch Model

In the computer based development process the sketch model is the point at which designers will turn to the computer tools at their disposal. The CAID system is used to generate a 3D virtual model which offers an alternative to 3D foam/card 'conventional' Sketch Models. However, several minor elements may still need to be modelled using the 'conventional' techniques as ergonomic decisions may not always be made effectively on the computer.

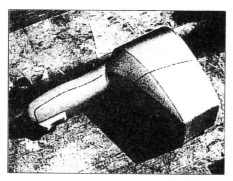

Fig 11 Trimmer sketch model in foam

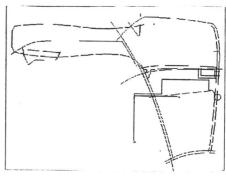

Fig 12 CAID construction lines for trimmer

The 'sketch' element of this type of model meant that even in the computer environment ease of generation is imperative. This 'ease of use' requirement has traditionally been the stumbling block for CAID systems. This extends to the ease of geometric interaction, not simply the user interface.

Taking the nylon line garden trimmer as an example of the CAID model building process the paper based sketch was taken as the starting point for computer based development of the trimmer, but such a complex form was impossible to generate as a single element or surface. The form was therefore broken down into simpler parts, which is a requirement of all CAD systems. These consisted of the hand grip being built from the three basic surfaces, and the cutter head was from four.

The design methodology of DeskArtes differs from traditional CAD systems by closely mirroring industrial design practice. In DeskArtes all but the most complex models are created using only two dimensional free-form curves which frees the designer from having to manipulate curves directly in three dimensions (a notoriously difficult process). The curves drawn in DeskArtes normally correlate closely with the curves the designer would draw in the 'conventional' environment of paper and pencil (figure 12).

To create each element of the nylon line trimmer the curves defining the projection (or silhouette) of a surface when viewed from an axial direction were drawn. These were combined with a second set of curves which defined the cross sectional shape of the surface at a given point between the original projection curves. Using this relatively small amount of information, DeskArtes generates a 3D surface. The shape of this surface was refined by creating and editing a further set of curves which described the shape of a surface when viewed from a second axial direction.

5.3   Computer Development Drawing

The distinction between the Sketch Model and the Development Drawing was even more blurred within the computer environment than using 'conventional' techniques. The ease and speed of computer based shading of 3D models meant that inevitably shading becomes as much of a tool to be used during the sketch modelling process as an end in itself.

To facilitate this shading a Silicon Graphics workstation was used as it has the capability to generate real-time shaded animations of the product. This ability is seen by some as a gimmick, but the experiences of the authors in this and other projects has shown it to be a valuable productivity aid.

5.4 Computer Rendering

Having generated and visualised a model it was a relatively simple step to generate a rendered image. The main processes involved were the assignment of material surface properties (colour and finish) to each component; to set up the scene which includes the setting of the 'camera' position and lighting of the object; and finally setting up the environment (floor and background) to enhance the final image.

Unlike the conventional rendering, the computer rendering was infinitely variable and could be viewed from whatever position required. This enabled modifications and even more choice to be made which would not be possible using 'conventional' rendering. Figures 8, 9 and 10 are ray-traced images generated from the computer model.

5.5 Physical Models

Whilst it took slightly longer to arrive at the final computer model it could be argued that the advantages over the 'conventional' techniques were apparent but not significant. However, it is the ability to create rapid prototypes that provides the tangible advantages to industrial design and NPD generally.

During the 'conventional' route a 'visual model' enables the assessment of form and basic ergonomics only. Later on the 'visual prototype' confirms engineering compliance and ergonomic acceptability. However, using rapid prototyping it is possible, with relatively little additional design effort, to effectively combine the two conventional physical model types.

Having converted the computer model into the appropriate stereolithography output the nylon line trimmer will be modelled as a rapid prototype as part of the second phase of this research exercise.

This has already been successfully completed with design and production of the ceramic coffee pot [1]. The stereolithography model was finished to mimic the product as it would appear 'in the shop' and used by marketing managers to show to potential customers. It therefore proved a valuable tool for the early stages of NPD.

6 CONCLUSIONS

Prior to the advent of dedicated CAID applications and rapid prototyping technologies, the nature of industrial design necessitated a methodology that remained virtually unchanged since the origins of the profession at the start of the century. Whilst being effective, this methodology favoured working practices whereby the industrial designer was almost detached from the overall NPD process. In many respects this was one of the reasons that favoured a largely consultancy-based profession.

As the hardware and software costs for CAID continue to decrease, along with more 'designer' friendly applications, a shift from the 'conventional' methodology is becoming increasingly feasible. Whilst the case studies undertaken have highlighted the benefits of a computer based methodology, it is felt that the advent of more accessible and cost effective rapid prototyping resources may ultimately be the development that proves to be a major incentive for adopting CAID. Once taken up the integration into concurrent engineering practices will be relatively straightforward, bringing its benefits to both the profession and clients.

The generation of a three dimensional computer model, to make use of rapid prototyping, opens up a variety of benefits within the process of NPD. These benefits have already been translated into reality by engineering and manufacturing disciplines, whereby computer models form the basis of concurrent engineering practice which allows discrete processes and disciplines to work in parallel to reduce time to market . These discrete processes are many and varied, but include analysis, simulation, manufacturing process and tooling development, early procurement of raw materials and components.

By bringing forward the engineering and ergonomic evaluation through the use of rapid prototype models, project time scales are reduced significantly. As complexity of form makes little difference to rapid prototyping costs, unlike 'hand built' conventional models, there may be financial gains. This would of course depend on the technique being employed, and whether post prototyping operations were required such as vacuum casting.

The product design exercises have indicated that the combination of CAID and rapid prototyping is now a viable alternative to conventional industrial design practices. CAID offers immediate advantages in the latter stages of the industrial design process by enabling a more rigorous analysis of form interactions and easily achieved photo-realistic renderings. Additionally, the use of rapid prototyping afforded by a three dimensional digital model extends the usefulness to the design/engineering interface. The introduction of CAID and rapid prototyping has wider implications since the digital model created can be used as the basis for the work of other downstream engineering functions.

7 REFERENCE

Cheshire, D.G., Harrison, D.K. and Wormald, P.W. (1994) 'Novel techniques for tableware development', in Dickens, P.M. (Ed) *Proceedings of the 3rd European Conference on Rapid Prototyping and Manufacturing*. Nottingham: University of Nottingham, pp 181-90.

Product variation simulation in design and pre-production manufacture

J HASTON and **N HAY**
Napier University, UK

Prototypes and computer models used in structural integrity and durability assessment have attributes that differ from those of the full-scale manufactured product they represent. Expected variations in the manufactured product, due to production tolerances for example, are not normally reflected in the prototypes and models, although the amount of variation allowed in production has a direct effect on the product's manufacturing cost and its structural integrity. Current research at Napier University uses static load testing and finite elements to investigate how structural integrity is affected by geometric variation from the manufacturing process.

INTRODUCTION

A vast range of modern computer-aided-design tools are now available to assist designers and engineers in their work. Recent developments have yielded a number of new technologies such as parametric features based solid modelling, geometric finite element analysis and solid free-form fabrication. These new product design, analysis and prototyping tools, although expensive, are now becoming well established.

Parametric solid modelling

Solid models of components and assemblies are efficiently represented in parametric form by the creation and combination of features whose dimensions are represented by parameters (1,2). When parameters are changed the effects are automatically rippled through the whole design, so that the solid model, finite element model and any assembly affected by the modifications are rapidly updated. The parameters themselves are often related through equations, which are used for scaling to ensure compatibility between mating components in an assembly or manufacturing set up.

Geometric finite element analysis

Recent developments in finite elements have yielded a new generation of analytical tools based on geometric p-type elements (3,4). These elements map the contour of the model's geometry and as a result fewer elements are needed than is the case with conventional h-type finite element modelling techniques. With the h-type of elements accuracy of the approximate solution is achieved by systematically refining the finite element mesh, in critical areas of the model, whilst using lower order polynomial equations for their solution. This is often a very time consuming process. However, analysis models based on the use of p-type elements enable the use of a coarser mesh. Although larger elements are used, accuracy of the solution is achieved by progressively increasing the polynomial order of the functions that describe their geometry, until a pre-defined level of convergence of selected solution parameters is achieved.. Time spent generating the mesh is reduced and confidence in the approximate solution is gained from the level of convergence attained. The pre-defined level of convergence is normally achieved without refining the finite element mesh.

Solid free-form fabrication

Physical models are also produced using automated techniques, including those solid free-form fabrication processes such as stereolithography, laminated object manufacturing, solid ground curing and selective laser sintering (5,6,7). These processes each accept geometric data directly from a solid modeller's database. The data is processed using specially formulated software and the output used to control the automated building of structures that faithfully represent the geometric characteristics of the original computer model. Extremely complex models are fabricated, often in a matter of hours rather than the weeks or months required when using conventional prototyping methods. Although solid free-form fabrication techniques offer advantages in some areas of product development, conventional manufacturing processes and materials are often employed for producing prototypes that are to be tested from a structural integrity point of view.

The design and analysis tools briefly discussed here provide essential information at an early stage in design and have reached a high level of maturity that gives confidence in the information generated by them. Nevertheless, physical testing either by laboratory simulation or service testing is used as the final test of structural integrity.

PRODUCT VARIATION IN FULL-SCALE MANUFACTURING

Complex assemblies such as automotive vehicles are manufactured using a broad range of production and assembly processes. During their design a certain amount of variation is allowed for by the application of tolerances to component and sub-assembly specifications (8,9). The tolerances are applied so that sufficient interchangeability between components that make up the product is achieved, enabling the use of high volume automated manufacturing techniques. The manufacturing processes used must also be capable of meeting these design tolerance specifications. If the tolerances specified are too tight the manufacturing process costs more to control and may not produce goods with the required precision. On the other hand, if the specified tolerances are too loose interchangeability could be jeopardised. In that case the end product may not assemble, be unreliable or be unsuitable for use. It is important therefore, to have a measure of the capability of the manufacturing system and to employ suitable tolerancing schemes that minimise production cost whilst ensuring sound product integrity.

Successful structural performance is based on stress analysis undertaken during design, with finite element modelling playing an increasingly important role. However, variations due to manufacturing and assembly tolerances, and other features such as notches and flaws, are often omitted in these models. Therefore prototypes currently used for assessment represent only a limited sample of an infinite number of geometric variations, so the consequences of variability in terms of material, manufacturing process, geometry, service loads, etc. are difficult to quantify (10).

VARIATION ANALYSIS IN DESIGN AND PRE-PRODUCTION MANUFACTURING

Commercial analysis packages based on the use of geometric p-type elements allow the modification of geometric features without a requirement for re-meshing the model. Therefore in addition to examining the state of stress for a given set of geometric and kinematic constraints, design sensitivity to changes in geometry can be studied. The following example involves a study of design sensitivity to variation in a spot welded structure, the finite element model of which is shown in Figure 1.

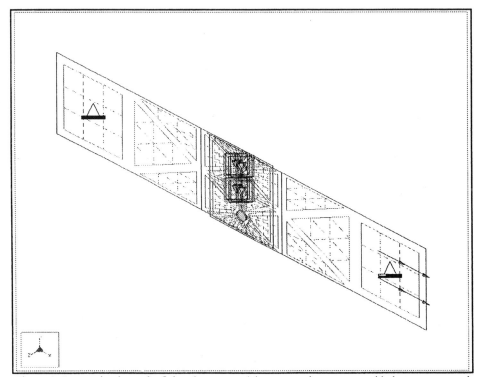

Figure 1. Isometric view of a finite element model, representing a spot welded structure, used for design sensitivity studies.

In this example the product consists of two thin steel strips fixed together by three spot welds, each of which tends to raise the stress locally when a tensile load is applied to the structure. If the spot welds have precisely equal diameters, are positioned with perfect accuracy and precision and if the geometry of the strips are perfectly aligned, then the area of stress concentration in the vicinity of each weld will be similar in size and shape as shown in Figure 2. Of course, it is unrealistic to imagine that these conditions can be satisfied in practice. A more realistic view acknowledges the presence of inevitable structural and operational imperfections, which means that the stress distribution present in consecutive units of the manufactured product will be different.

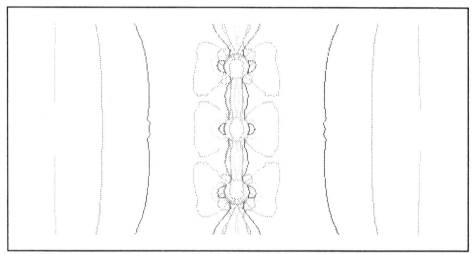

Figure 2. Contour map of the stress pattern obtained from a model where the weld detail is perfectly positioned.

To study the effect of variation in the position of the spot welds, before physical prototypes are available for testing, the whole structure, including weld detail, is modelled using finite elements. Whilst constructing the geometry for the analysis model, design variables are defined and associated with each spot weld. These design variables are then used to simulate changes in weld position during subsequent design studies. Results obtained from the design studies show how the magnitude and distribution of stresses in the structure change with respect to changes in weld position.

Figure 3 illustrates how the stress distribution in the vicinity of the welds is different when a simulated change in weld position is applied to the model. The change in magnitude of maximum stress in the model, with progressively larger deviations from the nominal weld positions, is shown in Figure 4. This example demonstrates the application of design variables to manufacturing variation analysis at the design stage. Information gained from the design study is then applied to the prototype test.

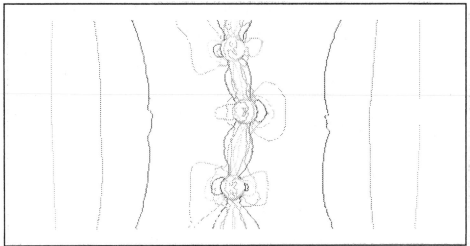

Figure 3. Contour map of the stress pattern obtained from a model where deviations in weld positions were simulated.

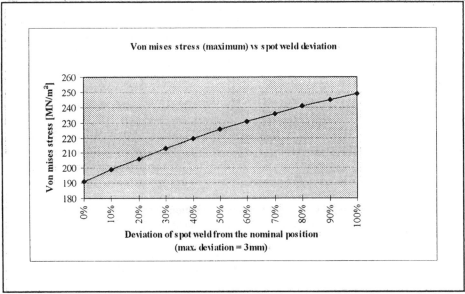

Figure 4. Graphical representation of stress data related to the position of spot weld design variables.

APPLICATION TO PHYSICAL PROTOTYPING FOR DURABILITY ASSESSMENT

Results obtained from the previous analysis show that the highest stresses are concentrated close to the edges of the welds and that the maximum stress increases as the welds deviate from their nominal positions. In this case, with a maximum deviation of 3mm, a 30% increase in stress is predicted with a proportionate reduction in fatigue life. This knowledge provides an opportunity to predict the variation in performance of the product in service. For example, studying the likely effects of geometric variations in stress sensitive areas during design analysis may indicate there are benefits to be gained by selectively controlling variations in the geometry of physical prototypes used for structural integrity and durability assessment. This could then provide important statistical data that gives a better formed impression of the spread of durability expected in large samples of the manufactured product.

CONCLUDING REMARKS AND FURTHER RESEARCH

Experimental tests are currently being undertaken to study the effects of geometric variations in laboratory test samples, to validate the solutions obtained from analysis models. Further work will include the application of statistical techniques to the prediction of service performance variation. It is expected that this work will serve to demonstrate the potential benefits of product variation simulation in design and pre-production manufacture. Analysing the effects of manufacturing variation at the design stage will enable a better prototyping strategy that gives a more realistic indication of how the manufactured product will perform in service. Physical prototypes used in laboratory simulation tests for example, will give a better understanding of the manufactured product when appropriate tolerances are applied during the testing procedure. The use of design variables in finite element modelling provides the means for achieving this.

REFERENCES

1.   Hay, N.C., Haston, J., Dodds, I.J. and Edwards, D., Product Optimisation Through Simultaneous Engineering: Proceedings of the 3rd International Conference on Product Optimisation for Integrity (Engineering Integrity Society), 1995.

2.   Haston, J. and Hay, N.C., Efficient Product Definition Through Computer-Aided-Prototyping, Eleventh National Conference on Manufacturing Research, 1995.

3.   Liu, A.F. and Gurbach, J.J., Application of a p-version Finite Element Code to Analysis of Cracks, AIAA Paper No. 93-1450, 34th Structures, Structural Dynamics and Materials Conference, 1993.

4.   Babuska, I. and Suri, M., "The p- and h-p versions of the Finite Element Method, an Overview", Computer methods in applied mechanics and engineering, Volume 80, pp 5 - 26, 1990.

5.   Wall, M.B., Ulrich, K.T. and Flowers, W.C., "Making Sense of Prototyping Technologies for Product Design", ASME 3rd International Conference on Design Theory and Methodology, 1991.

6.   Kochan, D., Solid Freeform Manufacturing - Possibilities and Restrictions, Computers in industry, Volume 20, pp 133 - 140, 1992.

7.   Waterman, N.A. (Ed), Rapid Prototyping in the USA, Report on the overseas science and technology expert mission to the USA, 1993.

8.   Mills, B., Variation Analysis Applied to Assembly Simulation, Assembly Automation, 1988, Vol. 8, pp 41 - 44.

9.   Kompella, M.S. and Bernhard, R.J., Measurement of the Statistical Variation of Structural-Acoustic Characteristics of Automotive Vehicles, SAE Noise and Vibration Conference and Exposition, 1993.

10.   Dabell, B. and Musiol, C., Simulation Testing and Durability Analysis: An Integrated Approach, Product Optimisation Through Simultaneous Engineering: Proceedings of the 3rd International Conference on Product Optimisation for Integrity (Engineering Integrity Society), 1995.

Case study of the application of LOM by Brook Hansen for the development of a housing design for small to medium electric motors

T J COOLE, D G CHESHIRE, and **D J NEWMAN**
Staffordshire University, UK
D K HARRISON
Glasgow Caledonian University, UK
A LOCKWOOD
Brook Hansen, Huddersfield, UK

ABSTRACT

This case study investigates the air flow over a casing or frame of an electric motor to improve the operating efficiency. Through redesigning of the frame it was possible to promote better thermal dissipation of the heat generated by the motor and reduce the windage loses and noise. It has been estimated that an improvements of 3% in the efficiency of these motors would generate an overall saving in the operating costs of approximately £1 billion per year in Europe alone.

The approach was to perform an analysis of the existing model to establish the current performance of the motor. A model of the new design was developed by rapid prototyping using the Laminated Object Manufacturing (LOM) system. As this was the company's first venture into rapid prototyping, computer models were generated in the initial instance and mathematical calculations were performed to test the accuracy of the proposed model. The paper concentrates on the methods of application for rapid prototyping into this type of industry and identifies the possible areas of potential problems.

1. INTRODUCTION

Brook Hansen is an engineering company which specialises in the development and production of small to medium size electric motors with a power range of 120W to 650kW. They have a distinguished record which goes back over the last 90 years. Heavy investment in new production techniques and design technology has enabled them to maintain a competitive edge with innovative and high quality product design. With energy efficiency being a major consideration in any improvements in motor design it was estimated that a 3% improvement in the efficiency of all electric motors in the EC would result in a saving of approximately £1 billion/year in consumed electricity.

In an effort to improve the efficiency of the power output to power input of the induction motors, the company identified several areas where losses occurred and identified improvements which could be made. It was identified that there were five key areas to energy efficiency. These are:-

1. Improve electromagnetic design for reduced copper loss.

2. Improve magnetic steel to give lower iron losses.

3. Improved thermal design to transfer heat more efficiently and at lower cost.

4. Improve aerodynamics to reduce windage losses and noise.

5. Improve manufacturing and quality control to reduce process related losses.

The pie chart shown in Chart 1 indicates the percentage of losses by each of these areas. The main heat sources inside the motor are copper losses on the stator and rotor windings and iron losses in the magnetic core. both these were reduced by the use of improved magnetic steels and lower resistance windings. In parallel, work to improve the thermal dissipation of the frame and particularly to reduce the volume of cooling air required (to reduce noise) was undertaken.

2. AREA OF APPLICATION

The first stage was to carry out an analysis of the performance of existing motors and to back this up by test results. From this work a thermal model of the motor was generated which allowed cooling optimisation calculations to be performed. These results looked promising but were unproved and needed to be backed up by tests on full size motors. Time was short and so the idea of using a rapid prototyping technique was born. Unfortunately the actual results can not be published in this paper due to the sensitivity of the information and its value to competitors.

To decide on the RP process that would be most suitable for this type of application, a range of systems were looked at from Stereolithography (SLA), Selective Laser Sintering (SLS), Laminated Object Manufacturing (LOM), and Fused Deposition Modelling (FDM). The model had to perform to specific constraints that would allow the creation of tooling for an accurate working part in aluminium. The other constraints were cost of the model production and the size of the component that was being made.

The motor for testing was the WD132. This was chosen as a mid range motor from which it was hoped that characteristics could be extrapolated to other motors in the range. The size of the model that was required in this instance, 280 mm length x 264 mm diameter was the controlling factor in the final selection of the most appropriate Rapid Prototyping process. The LOM system had the capacity to build the model in the correct build orientation and UMAK offered a bureau service.

The 3D model was developed on the I.D.E/A.S. CAD system (Figure 1), and from this an STL file was generated. The .STL file had a total of 95,000 facets to maintain the surface definition of the model. Through the use of Finite Element Analysis, the model was tested for the loads that would be exerted on the motor once the unit was in operation. Figure 2 shows the graphical results of the end casing of the model using cyclic loading that would occur in the operation of the final product. These result highlighted the stresses areas that would occur in the model and showed that the worst stresses would occur around the locking lug.

As the model was to be used to create an aluminium casting of the component, that could be used for full functional testing, the prototype model had to be post processed to reproduce this model in the production material (aluminium). Some of the tests could be performed on the RP model however it was felt that more accurate results could be achieved if a full working model. In this instance investment casting was the post process used to generate the test prototype. This process is described in section 3.

The LOM model was made by UMAK Limited on the Helisys 1015 machine (Figure 3) as two separate sections and nested together in the build envelope for the main body. The two endshields were made separately. The total build time for this unit was 95 hours. On top of this was the breakout and finishing which took a further 10 hours. The build process was stopped several times to remove waste material from any areas that it was felt might cause problems in the finished model. This was made possible by the flexibility of the system being able to assess the status at any stage in the job cycle. Table 1 gives some of the build criteria which were set for the operation.

3. THE CASTING

The casting process was performed by Deritend Aluminium Castings Limited of Worcester. The surface of the LOM model was first sanded and then sealed with a cellulose sealant to prevent de-lamination in the burnout process. Once sealed a thin layer of ceramic investment material was used to coat the model. While the slurry was still wet a fine refractory sand was used to coat the shell [1]. This was to ensure that fine details are reproduced. The model was then repeatedly dipped in slurry to give any even coating to the whole of the surface. The ceramic shell was left to set and dry before being put through an autoclave cycle to remove the wax gating attached to the pattern.

The ceramic shell was fired at a temperature of 550^0C to remove the LOM model from within the shell. As this temperature is fairly low the model is burnt out before the thermal expansion forces are great enough to crack the shell. This process leaves a residue of about 5-10% ash in the shell which was then eliminated with either compressed air or fluid rinsing [1].

The shell was then preheated to ensure than an even flow was achieved to all detailed sections of the mould. The preheat temperature of 660^0C which is to the melt temperature of the case alloy used, in this case Al-25 [2]. The pouring was performed using the gravity feed method with no assistance from pressure or centrifugal feeding.

The aluminium casting was then broken-out from the shell. This process can accurately reproduce this sections down to 1 mm while maintaining a high level of dimensional accuracy. Using this process provided an efficient method for creating a working prototype

4. THE MODEL

Once the casting was produced inspection showed there were several minor discrepancies which required attention (Figure 4).

The fins on the outer casing of the motor were thinner than expected. The cause of this was that when the model was first designed by the Modeller, allowance for the shrinkage of the aluminium was not taken into consideration. The shrinkage rate for aluminium is 0.8%. The reduction in thickness was not considered to be a major problem as the heat loss was only affected by a minimal amount due to his [2]. A second prototype which properly took account of shrinkage effects gave dimensions very close to drawings. The laminated layers that go to make up the LOM model were still apparent on the surface of the cast prototypes. A small amount of machining and finishing of the casting was required to bring it to a satisfactory condition for testing the improved performance which was expected from the theoretical results.

5. RESULTS

The resulting component was extensively tested to prove the design concepts. By generating the model through this approach the working prototype could be generated for as little as £6270[3]. This meant that any design changes that had to be made could be incorporated into the final tooling without having gone to the expense of producing the tooling in the initial instance.

Although the model was produced at a far lower cost than could have been done by any other method. It is important to look at the real value of RP in the context of where the benefits are actually achieved. This model allowed the company to develop a test motor in a considerably shorter period of time (see table 2) than would have been possible by the traditional route. It was also done at far lower cost. This allowed valuable testing to be done at a realistic cost. Emphasis should be placed on the fact that design alternations can not always be made on the conventional technique without retooling. The final production tooling still had to be generated at the same cost using traditional techniques. However, the savings come in reduced re-engineering by eliminating faults in the early stage of the design cycle prior to expensive tooling. The cost saving, had retooling been required, could have been as high as £100,000 plus the time delay of up to 24 weeks (3).

This coupled with the improvement in the manufacturing quality control of the assembly and production of the motor components, enabled the target of a 3% increase in efficiency achievable. The main area of progress was in the alteration of the air flow this allowed a better dispersion of the heat from the body improving the efficiency by about 1% and contributed substantially to the overall results of the research. The testing would only been carried out after full tooling and changes would have been very expensive.

6. CONCLUSION

These Brook Hansen 'W' range motors have been the first product range to benefit from this research. The 3% improvement in the efficiency was achieved with little or no extra cost in the production process. Through the introduction of the LOM process into the development cycle accurate design of a range of other product parameters was achieved which could not be proved without practical testing. The efficiency has been increased and the noise reduced without adversely affecting other performance criteria such as starting torque, starting current and power factor. The running performance was good over a wide range of load conditions. The research has resulted in a range of new products designed to ensure the competitiveness in the world market and has proven that RP has a significant impact on the effectiveness of design in new engineering products. Figure 4 shows both the LOM model and the aluminium casting made from this model.

This paper is some of the ongoing results of the research that is being carried out by Brook Hansen. The follow on research is predominantly involved with improvements in iron quality and copper content of the motor.

7. REFERENCES

1. Dunham G.D. "Rapid Prototyping Through Investment Casting" Proceedings in the 52nd Annual Technical Conference. San Francisco, USA 1994 pp.1094-1097.

2. Walters D.G.; Williams I J. "Energy Efficiency of Electric Motors - By Design" Report on the 'W' Research Programme, Sponsored by the EEO 1994

3. Report from UMAK on the work carried out by themselves for the production of the LOM model for Brook Hansen. "The Brook Hansen 7.5kW World Series Electric Motor" 1995.

Tables

Criteria	
Name of Component	Motor frame and End shields
No. of Components	3
Size of Component	Frame 260 x 280 2 x End Shields 250 x 60
Resolution of CAD Model	0.2mm
Build Time	95 hours
File Size	95,000 Facets
Breakout Time	10 Hours
Dimensional Accuracy	±0.25 mm
Machine Used	Helisys LOM 1015-005
Paper Thickness	0.1 mm

Table 1. The Criteria used for the Prototype Production

RAPID PROTOTYPING METHOD			CONVENTIONAL TECHNIQUES		
Item	Cost	Time	Item	Cost	Time
CAD Model Generation	£10,000	3 weeks	CAD Model Generation	£10,000	3 Weeks
S.T.L Files	£1,200	1 Day*	Pattern Frame of Body	£60,000	24 Weeks
LOM Model	£3,240	4 Day	Pattern End Shields	£45,000	19 Weeks
Casting	£2,100	2 Weeks			
Total	£16,000	6 Weeks		£115,000	27 Weeks

* This represents the time allowed for correcting holes and flaws in the surface.[SMG1]

Table 2. The comparison between the cost and time saving of the Rapid Prototyping Technique and the Conventional Technique

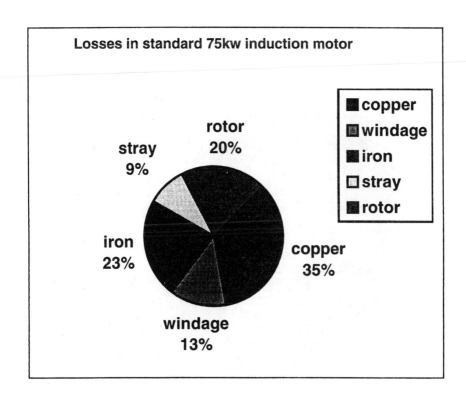

Chart 1. The breakdown of where losses occur and their percentage effect on the efficiency of a 75kw motor

(Produced by Brook Hansen in their promotional data)

TOP WORK

Figure 1. <u>The CAD model of the motor housing that was supplied in .STL for the generation of the LOM model.</u>

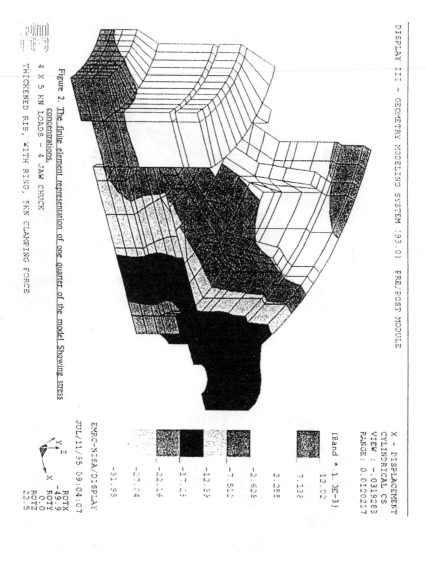

DISPLAY III - GEOMETRY MODELING SYSTEM (93.0)   FRE/POST MODULE

X - DISPLACEMENT
CYLINDRICAL CS
VIEW : -.0319283
RANGE: 0.0120217

(Band * 1.0E-3)

12.02

7.138

2.255

-2.628

-7.512

-12.39

-17.28

-22.16

-27.04

-31.93

EMRC-NISA/DISPLAY
JUL/11/95 09:04:07

ROTX   -49.9
ROTY   0.0
ROTZ   22.5

Figure 2. The finite element representation of one quarter of the model Showing stress concentrations.

4 X 5 KN LOADS - 4 JAW CHUCK

THICKENED RIB, WITH RING, 5KN CLAMPING FORCE

27

Figure 3. The LOM Model With The Two End Plates

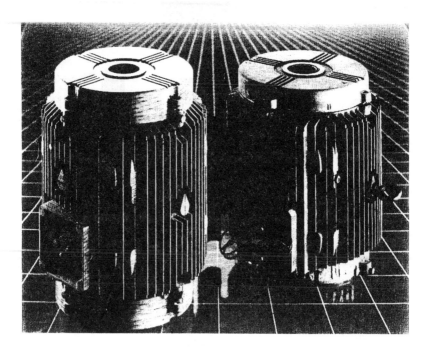

Figure 4. The Completed Model and The Aluminium Casting.

Rapid prototyping in the development of instrumentation for minimal access surgery – a case study

B PEARCE and **S S MORAN**
Surgical Innovations Limited, Leeds, UK

SYNOPSIS

Surgical Innovations is a company designing and manufacturing instruments for use in Minimal Access Surgery (MAS) — more commonly known as 'keyhole' surgery. This is a fast moving, highly technical market requiring well engineered, high quality products. In order to maintain its competitive edge and enhance its differentiation Surgical Innovations is utilising Rapid Prototyping (RP) in a number of ways:—

- Mechanical Prototyping in a dedicated research and development facility, enabling concept evaluation and initial clinical trials.
- 3D solid modelling used to generate Stereolithography (SLA) models, facilitating ergonomic and aesthetic appraisal, advanced clinical trials and market assessment studies.
- 3D solid modelling used to generate data suitable for direct transmission to mould tool makers, reducing interpretation errors and mould lead time.

Surgical Innovations has successfully utilised this technology in the following projects:—
- *EndoFlex TA*
  Redesign of the EndoFlex Retractor System — instrumentation which simulates the surgeon's hand — incorporating 'take-apart' features designed to ease cleaning, improve ergonomics and aesthetics.
- *FastClamp*
  Design of an operating room table clamping system for orientating and holding surgical instruments within the body thus freeing-up operating room staff.

The company has realised notable benefits through the use of RP:—
- Ability to create and clinically test new products in a short space of time.
- Overall reduction of 'time to market', therefore, enhancing sales opportunity.
- Significant reduction in tooling costs — key in a small to medium enterprise (SME).

1. INTRODUCTION

Over the last few years there has been a surgical revolution. *Minimally Invasive Surgery (MIS)* [1], or the term used throughout this paper — *Minimal Access Surgery (MAS)* [2], also known colloquially as 'keyhole surgery' — describes procedures that are performed through a series of small incisions, thus greatly reducing wound trauma. The surgeon manipulates small diameter instruments through ports, or trocars, which are inserted through the incisions. Hence, the inherent tissue trauma associated with dissection is greatly reduced. Since the body effectively remains closed the problems of cooling, desiccation, handling and forced retraction are almost eradicated. The traumatic assault normally experienced by the patient is greatly reduced enabling the patient to make a much speedier recovery with significantly reduced convalescence. Post-operative hospital costs are greatly reduced and the patients, their employers and the economy benefits from their early return to work.

2. BACKGROUND — THE TECHNOLOGY

In a survey, conducted in 1992 [3], 300 surgeons from the Society of American Gastrointestinal and Endoscopic Surgeons [SAGES], an association which is a leading advocate of the use of MAS in the United States, were asked a range of questions about their use of MAS. When asked to state their major motivations for utilising MAS their responses were as shown below in Figure 1:

- **Reduced patient convalescence**

- **Reduced post-operative complications**

- **Patient's request for MAS**

Figure 1 Surgeons' Motivations for Using MAS

However, there are difficulties associated with this type of surgery. The surgeon is forced by the nature of MAS to operate remotely from the surgical field. Unlike open surgery, where the interface between the doctor and patient is a small hand held instrument, minimally invasive techniques usually require long instruments with a shaft length in the range of 300mm (12"). The operation is observed through a series of optical and electronic couples, ultimately producing a two-dimensional image on a video monitor. There are problems with hand-eye co-ordination, depth perception, and interference of the tactile sensation.

'In general procedures performed using MIS require more technical expertise and are slower to perform...............................' [4].

3. BACKGROUND — THE COMPANY

It is in the light of the problems and opportunities described in Secion 2 that Surgical Innovations Limited was founded four years ago. The company has established a global marketing infrastructure and has grown through sales of products based on its novel flexible, segmented technology, known as EndoFlex (see Section 5).

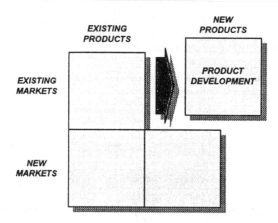

Figure 2 Surgical Innovations' Growth Strategy [5]

The company now has a significant growth opportunity by developing products to meet its customers requirements. Surgical Innovations' choice of growth strategy is summarised in Figure 2.

Surgical Innovations employs fourteen people, four of whom are permanently employed in the area of product development, and therefore can be classified as a small to medium enterprise [SME].

4. WHAT IS THE OF ROLE RAPID PROTOTYPING ?

The market for MAS products is fast moving, highly technical and requires well engineered, quality products. In order to maintain its competitive edge and enhance its differentiation it is necessary for Surgical Innovations to generate a constant stream of innovative products. Furthermore, in a dynamic market place it is essential to compress the product development cycle in order to take advantage of growth, for example, in a particular endoscopic procedure. The company is attempting to meet this challenge by employing its own brand of rapid prototyping.

Rapid Prototyping [RP] has been defined as:

'A technique in which physical models are fully created from materials, provided in various forms, completely under the control of model data created within a computer aided design environment.' [6]

Surgical Innovations version of RP can be defined somewhat differently:

'The ability to create a working prototype that fulfils the initial aim of the project in the shortest possible time. The prototype should be such that it can be adapted and altered at any stage and is capable of use in a clinical environment.'

In order to justify this re-definition it is important to understand the *idea generation mechanism* and subsequent *prototype generation mechanism*.

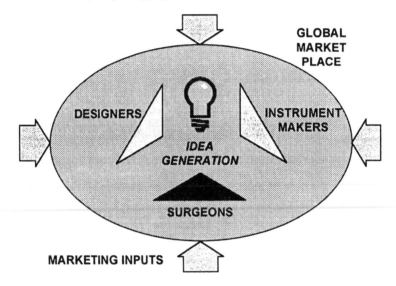

Figure 3 The Idea Generation Mechanism

Figure 3 illustrates the key elements of the *idea generation mechanism*. Surgical Innovations is a market oriented business and is focused on meeting surgeons needs. The successful implementation of this mechanism is dependent upon the merging of skills and experience of designers, instrument makers and surgeons.

The *idea* could be assimilated from internal sources or external sources such as surgeons or market experts. Once the idea has been formalised it is fed into the *prototype generation mechanism* shown in Figure 4.

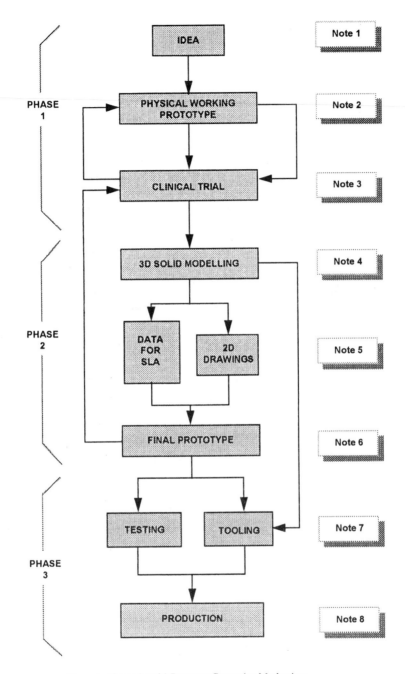

Figure 4   Phases 1 and 2 Prototype Generation Mechanism
Phases 1, 2 and 3 Product Development Cycle

Note 1:	The idea emerging from the *Idea Generation Mechanism* [Figure 3] is fed into the overall *Product Development Cycle*.
Note 2:	The design team, summarised in Figure 3, devises a design brief. The design brief is then realised by the instrument makers using materials such as brass for handles and components outside of the body and stainless steel and engineering plastics for components inside of the body.
Note 3:	The surgeons are now able to use the prototype in surgery. Although the device may be of a basic construction, emphasis is placed on key factors such as tactile feedback. This allows the rapid evaluation of the concept and feedback of any necessary improvements. Since the prototypes are manufactured from readily workable materials adding or subtracting features, or even starting afresh, can be effected in terms of hours and days.
Note 4:	Once the concept has been verified and the key features of the prototype finalised, 3D solid modelling of the device can begin. Arguably this is sequential engineering but constant involvement of the design team ensures a high level of knowledge exchange and a short learning curve for the design engineer responsible for solid modelling.
Note 5:	Typically machined metallic components are combined with injection moulding in the final product. The computerised solid model is used to generate 2D engineering drawings and data for SLA components.
Note 6:	The data is appropriately converted into physical components and these are assembled into advanced prototypes. The process discussed in Note 3 is repeated and the prototype refined where necessary. Any significant changes at this stage will mean that the 3D solid model must be re-created and the following stages repeated.
Note 7:	When the prototype has been approved data can be transferred directly to the relevant parties, e.g., tool makers. During this phase in-house, mechanical testing of the prototype product takes place. Any changes deemed necessary as a result of testing must be fed back into the process at the beginning of Phase 2.
Note 8:	On completion of the tooling the components are manufactured and assembled.

Figure 4 Accompanying Notes
Phases 1 and 2 Prototype Generation Mechanism
Phases 1, 2 and 3 Product Development Cycle

The process shown in Figure 4 emphasises the differences in the definitions stated at the beginning of this chapter. Phase 1 of the process has significantly less or no reliance on Computer Aided Design [CAD] and places great emphasis on the production of physical models capable of use in surgery. This is the essence of Surgical Innovations' brand of RP.

Surgical Innovations has found that once Phase 2 of the RP process has been entered costs can quickly escalate with the involvement of external agencies, such as CAD bureaux and SLA machine time suppliers. With each design change requiring another expensive model the costs of using modern RP techniques can begin to have an adverse effect on the design.

The benefits of pursuing RP, such as reduced lead times and reductions in tooling costs, are well documented and have been capitalised upon by Surgical Innovations during Phase 3. However, until there is a reduction in the costs of producing prototypes using, for example SLA modelling, and an improvement in the mechanical properties of the available materials this is not a route that can be undertaken during the early stages of projects.

5. RAPID PROTOTYPING IN ACTION

5.1. 'Take-Apart' EndoFlex Retraction System

EndoFlex technology enables an instrument to be inserted into the body through a standard 5.5mm port in a straightened condition and then actuated into a rigid pre-defined shape. Subsequently, the instrument can be straightened again and removed from the operating field. The flexible section of the instrument is constructed from a range of inter-linking segments as illustrated in Figure 5 and can be used to produce a range of shapes which are suitable for retraction and mobilisation of large organs and tissues structures within the body. These instruments effectively replicate the surgeons hand during MAS. Typical examples of the shapes are shown in Figure 6.

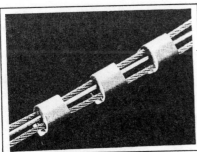
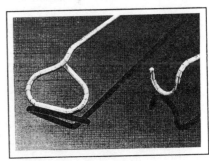

Figure 5 Segment Technology   Figure 6 Curved Instruments

A perceived disadvantage of this product is the difficulty associated with cleaning the segmented section.

Although Figure 5 shows the segments spaced apart it was not possible to achieve this condition with the original instrument. It had long been known that the ability to space the segments apart and gain access to the inside of the main shaft of the instrument would facilitate ease of cleaning. As a result a 'take-apart' prototype was developed using many of the existing components.

Figure 7 Original Handle

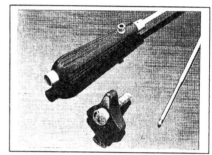

Figure 8 Physical 'Take-Apart' Prototype

Figure 7 shows the handle of the original instrument, whereas Figure 8 shows the integration of the 'take-apart' features. For technical and marketing reasons it was deemed necessary to re-engineer the handle. The main project goals were:—

- House the actuating mechanism within the handle.
- Improve the ergonomic characteristics.
- Improve the overall aesthetic.

Once the mechanical feasibility had been investigated (using the instrument shown in Figure 8) traditional model making techniques were employed.

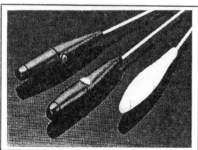

Figure 9 Different Modelling Techniques

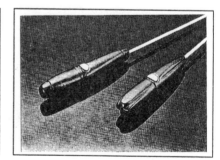

Figure 10 Wood Block Models

Figure 9 shows the three different modelling techniques used. On the right, the light coloured handle, is a foam model. The use of foam allows the designer to create many ideas quickly and cost effectively but the models have poor definition and weight relationship. However, they enable the range of possible solutions to be significantly narrowed and form the basis for the production of wood block models. (See Figure 10)

This allowed the designer to add colour and realistic details. These models formed the basis of CAD 3-Dimensional Solid Modelling.

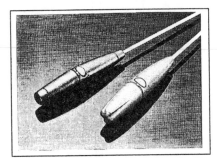

Using Pro/Engineer solid modelling system, available from Parametric Technology Corporation, it was possible to model the handle shape, all the associated mechanical parts, undertake interference checking of the assembly and analysis of the kinematics of the mechanism.

Figure 11 SLA Models

In parallel 2D drawings were issued for manufacture of the metallic components and data generated for the manufacture of Stereolithography (SLA) models. Once the machined parts were finished it was possible to assemble them with the SLA models and produce working prototypes which resembled finished product (see Figure 11). Even at this late stage two possible solutions were pursued.

The prototypes were evaluated by members of the company's clinical advisory board and exhibited at international congresses. One of the solutions was chosen but was deemed too heavy and bulky. The parametric nature of the CAD model allowed significant changes in geometry to be implemented quickly and at low cost. The revised model was again used to produce SLA components and the results are shown in Figures 12 and 13.

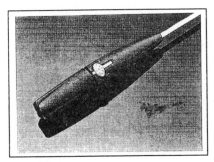 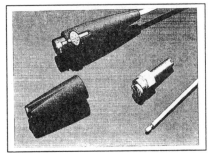

Figure 12
Final 'Take-Apart' Handle Prototype shown assembled

Figure 13
Final 'Take-Apart' Handle Prototype shown dismantled

In order to further differentiate the product it was decided to add a second colour to the end cap section of the handle using a technique known as 2-shot moulding. Figure 14a, 14b and 14c show the three stages involved.

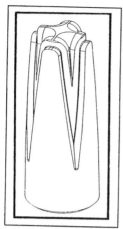 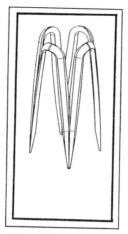 

Figure 14a
Material is removed from the original shape to form the first shot of the injection moulding process. The first shot is then enlarged to provide a step between the first and second shots — this is the 'shut-off' between the two injections.

Figure 14b
The material removed from the original is saved to form the second shot. The second shot provides the contrasting, coloured highlight.

Figure 14c
The first and second shots are combined to provide a single, two coloured component. The final result is a highly differentiated, aesthetically 'striking' component.

Once all aspects of the design were finalised digital information was transferred directly to the mould tool makers. This removes errors associated with the interpretation of complex surfaces and reduces tool design costs.

As a small company it is difficult to quote 'record-breaking' timescales for taking products to market as many different projects have to be pursued simultaneously and resources allocated accordingly. Nonetheless, it is the view of the authors that using the type of RP described in this document reduced the time from conception to completion by at least 40%, compared to conventional product development techniques.

5.2. Other Successful Applications of RP

EndoFlex Forceps

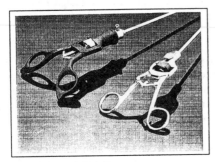

On the RHS of Figure 15 is shown an early working prototype of the EndoFlex forceps handle manufactured from brass. This device was used many times during surgery and was the basis of the SLA prototype that yielded the production model, shown on the left.

Figure 15 EndoFlex Forceps

FastClamp

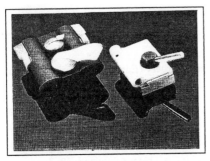

The devices shown in Figure 16 are designed to locate on the side of an operating room table where they are combined with an articulating arm to hold surgical instruments without the need for human participation.
The basic prototype is shown on the right of Figure 16 and again the device formed the basis for the advanced prototype shown on the left.

Figure 16 FastClamp

The knobs on the final prototype were moulded in Polyurethane from silicon rubber moulds — SLA models were used as patterns.

6. CONCLUSION

Surgical Innovations is making extensive use of rapid prototyping techniques.

More importantly Surgical Innovations is using its own brand of RP to design, build and test ideas in a clinical environment — the time scales are short and the nature of product evolution dramatic.

There is no doubt that the projects discussed in this document could not have been implemented with such a high degree of success and within their respective timescales if such techniques as Stereolithography had not been employed. Furthermore, the cost savings in terms of reduced tool design time and the speedier generation of cash-flows, due to reduced time-to-market, have been capitalised upon.

However, in the design of surgical devices it can be argued a broader definition of RP is required. The definition should be centred on rapidly creating prototypes capable of use in a clinical environment, rather than emphasis on a computer aided design.

In conclusion, both approaches to RP play an important role in the future development and success of Surgical Innovations.

7. REFERENCES

1. Wickham J, Fitzpatrick M, Minimally Invasive Surgery (Editorial), British Journal of Surgery, Vol.: 77, 1990

2. Cuschieri A, Buess G, Périssat J, (Editors), Operative Manual of Endoscopic Surgery, Springer Verlag, 1992, p. 10

3. Moran S S, A Survey of the Growth in Minimal Invasive Surgery in the United States, [Unpublished] Kingston University, Kingston-upon-Thames, England, 1992

4. Cuschieri A, Buess G, Périssat J, (Editors), Operative Manual of Endoscopic Surgery, Springer Verlag, 1992, p. 12

5. Ansoff H I, Strategies for Diversification, Harvard Business Review, Sept./Oct. 1957, p. 113-124

6. Medland A J, Rapid Prototyping Development.....the role of physical models in design, Rapid Prototyping Conference, Missenden Abbey, Nov. 1994, Buckinghamshire College

Rapid prototyping of manufacture of scale model aircraft components

D THOMAS
Nottingham Trent University, UK
J MOORE
Axiomatic Technology Limited, Lowdham, UK
MOFFATT
Sequoia Systems Limited, Wraysbury, UK

Synopsis

Rapid prototyping typically conjures up the impression of very expensive specialised equipment used in a novel way to realise three dimensional objects. The authors present an alternative and in some ways more conventional approach based upon the use of an innovative cartesian robot cutting machine in which the control systems is tightly integrated with a comprehensive CAD system. The Pacer Compact machine used was originally developed for sign manufacturing, but the ingenius integration of the CAD system with the control system produced by Axiomatic Technology, has realised a cost effective prototype and manufacturing capability for 2D and simple 3D components.

1 INTRODUCTION

In the minds of many, rapid prototyping generates an image of graphic solid modelling systems linked to sophisticated stereo lithography machines producing three dimensional objects using exotic technology. The rapid realisation of prototypes does not however need to rely on such technology and this paper describes work undertaken by the

The computer interface integrated directly with the machine, which can take material up to 800x900mm and 75mm thick.

41

authors to develop a rapid prototyping system suitable for producing components used in building miniature flying model aircraft.

Design is the key to realisation of any artefact, and a successful design facility allows the designer to exercise creative skills whilst supported by appropriate technical tools. Conventional CAD is readily available, certainly for 2D design and increasingly so for 3D work. However, it is usual for the CAD function to be considered distinct from the production operation. The approach adopted by the authors tightly couples the design and production operations through integration of the machine control within the CAD environment of a PC based design workstation. The resultant direct control of a high specification 3 axis CNC cartesian robot cutting machine manufactured by Pacer Systems Ltd (Ref. 3) provides a highly efficient and cost effective prototype design and manufacturing system capable of making a wide range of parts. The integration of a desktop scanner for design capture, further increases efficiency.

The power of available low cost personal computers is now such that extensive design data processing can be performed in a very short time. Provision of a suitable display interface, graphically displaying the activity, makes even the heavier design data processing tasks acceptable in an interactive design environment. In many cases, CAD data post-processing can be made to appear virtually transparent to the user by means of suitably automated functionality. At the same time, optional flexibility can be built in to allow custom configuration for specific requirements. The system produced by the authors allows the storage and selection of a wide range of different set ups to suit the variety of work undertaken.

2. THE CUTTING EDGE

A useful rapid prototyping tool should not interfere with the prototype design intent, be able therefore to cut and shape a wide range of materials to suit the prototype purpose. This should take into account not only working parameters, but also those relevant for future manufacturing.

Our machine can cut materials as varied as glass, stainless steel, wood, foam plastics and composite materials in a range of thicknesses up to 75mm depending on the cutter length. This will be related to the tool diameter which is dictated by the delicacy of the work involved.

The software allows for the bulk removal of material inside and outside a given shape, with a convenient way of accurately specifying the cutting depth. This enables relatively complex shapes to be cut from solid, however where this falls short of the design requirement, layers of material can be cut (at any thickness up to 75mm), and laminated using accurate registration techniques.

The system achieves all this using three-axis control of a precision-engineered cutting head mounted within a frame designed to achieve machining tolerances close to those of milling machines, but with greater flexibility of material handling.

Practical rapid prototyping is achieved because this flexibility and precision is linked directly to the CAD vector drawings with all the machine control parameters critical for successful operation transparent to the operator. Not only is the action of the machine rapid in itself, the speed and ease of use encourages testing of alternative ideas and improves the quality of development as well as reducing the time to achieve it.

Time to produce a prototype will depend on the operator establishing the best way to realise the intended shapes, and the integrated CAD software minimises this. The actual machine cutting speed is variable from 0.01 to 4 metres per minute to suit the material and cutting tool combination. Similarly, cutting depth can be varied to suit the material. For instance, some plastics such as ABS or polycarbonate melt easily if too much depth of material is cut in one pass. Our software allows the machine to automatically cut the material in a number of passes, thus keeping the workload light. Together with feed rate control, the best surface finish can be achieved on the work. Note that depth and feed rate can be altered 'on line' as experience dictates.

3. THE DESIGN PROCESS

Traditional component design for production on a CNC machine generally involves extensive off-line preparation, probably using a CAD workstation. CAM software, or postprocessing software is used to prepare the design data for transfer to the CNC machine controller which actually controls the axis movements of the cutting machine. This process is time consuming and requires a range of high technical skills in addition to the creative skills of the designer.

The CAD facilities integrated within the system implemented by the authors, incorporate

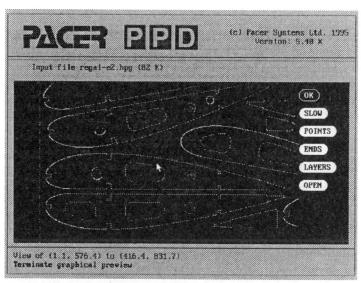

Part of the wing-rib design being processed by the PPD software on the PC

all the technical design facilities which might be expected within a modern design workstation. Using a modified version of the popular RoboCAD software, an elegant and rapid graphical interface is provided, together with an easy to use graphical filing system. The beauty of such a system is that despite its comprehensive specification, it may be used at any level of sophistication by the designer and in general the designer is unaware of the processes being invoked as the graphical image of the design is created on the screen.

This is in marked contrast to the rather more basic geometric or numerical specification of components to be produced by conventional methods. Such methods are not only tedious, but highly error prone. Indeed, it is frequently necessary to refine designs after components have been produced. In the main this can be attributed to the very low level of user interaction, not unlike machine code or assembler programming of computers which whilst still associated with some specialist aspects of programming is hardly the most efficient way of harnessing the power of modern computing machines.

A drawback however of the 'higher level' design is that the user has little direct control over data structures and element ordering within the data files produced. Indeed, frequently display features take precedence in order to provide the user with a fast responsive interactive design facility. Many CAD systems do allow some sorting of the output data either to reduce pen changing time or to increase the efficiency of plotting by plotting continuous contours. However, the basic facilities of the CAD system do not normally satisfy all the requirements of cutter path generation and sequence control.

For this reason we have developed customised design data processing software (internally called PPD). Although this software contains a great deal of functionality, for straightforward work its operation can be automatic and virtually invisible to the user. The functionality of PPD is described in section 6 of this paper.

4. SCANNED INPUT

Many shapes used in aircraft construction involve the specification of complex curves. Naturally, such curves can be either digitised or geometrically constructed. Both of the processes, though precise are time consuming and hardly consistent with rapid prototyping. Our preferred approach for rapid component prototyping, where the artwork is within the required dimensional tolerances, is to use a flatbed scanner with suitable vectorisation software, such as AxiSCAN.

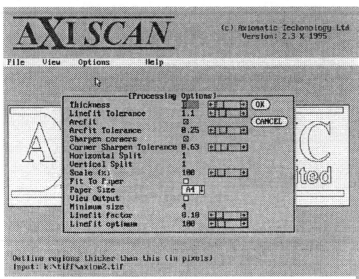

The scanner produces a raster image of the design and in order that it can be machined the raster must be converted to a vector format. Many vectorisation systems are available in the market place, but most either produce too much output in order to accurately represent the shape, or they produce unacceptable shape quality for the job

AxiSCAN is used on the PC to scan in a drawing and vectorise it. The resulting file can be edited with the CAD package, and sent to the machine for cutting.

in hand. Excessive data is a problem, both in terms of smoothness of cutter path leading to poor quality cutting and also in terms of the mathematical burden placed on the processor in computing the cutter compensation of the tool path.

AxiSCAN, our vectorisation software, provides accurate contour outlines of the shapes to be cut, using a combination of centre line and outline finding techniques. Nevertheless, in general even with good quality artwork problems may arise in the definition of smooth aerodynamic curves, even though the software generates curves as well as lines. These problems are satisfactorily resolved through the application of algorithms arising from research work of Poliakoff (Refs. 1 & 2) of The Nottingham Trent University in respect of curve fairing and data reduction. The resultant curves are thus accurate in shape to the artwork, aerodynamically smooth and consequently result in rapid high quality machine operation.

5. MACHINE CONTROL REQUIREMENTS

Most components, including many three dimensional components are produced from rigid sheet materials. As weight is important in aircraft, balsa is frequently used, often in combination with various grades of plywood and aluminium In some cases synthetic foam based materials may be required. Even with the use of these materials, most components require lightening holes.

Consistent high quality cutting on these materials naturally requires the selection of both the appropriate cutting tool and an appropriate means of clamping the material. The machine is equipped with a variable frequency inverter driven spindle allowing speeds up to 18 000 rpm. A range of cutters with diameters typically ranging from 0.4 mm to 6.0 mm diameter is used. The machine feed control based on work of Nottingham Trent University (Refs. 7, 8 & 9), results in smooth continuous path motion at near constant surface speed. This means that the smallest cutters can be used on harder materials without risk of breakage through erratic motion. It also means that a consistent high quality surface finish can be obtained on the workpiece.

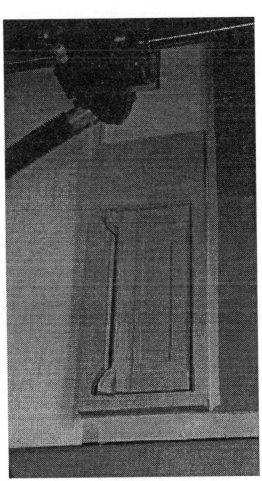

Part of a model bridge being cut on the machine

Clamping the work in important. Rapid production of prototypes rules out the use of special fixtures and mechanical clamps, which in any case would tend to damage the delicate materials employed. Provision of a powerful vacuum system over the whole bed provides a partial solution, quite satisfactory for the larger components, but smaller components are not effectively held and there is the possibility of damage on finishing the cut. This problem is overcome by an automatic feature (with manual override) of the software, whereby holding tabs are placed sufficient to hold the component in the sheet material. This not only makes for clean cutting of the smallest components - including realisation of components which could not otherwise be cut - but it also facilitates handling of sets of cut components and avoids damage.

The machine controller is implemented through a custom designed autonomous processor card connected to the bus of the design workstation. Using a dual port RAM interface to ensure highest performance, the controller communicates with the host CAD environment through a sophisticated bidirectional software driver. This allows both the passing of sequential data files of machining information and direct interactive controls. The large memory capacity of the controller and its in-built processing power means that it imposes very little processing overhead on the CAD host.

6. BRIDGING THE GAP

The elements of the system described above will be recognisable to many as essential elements in the CADCAM process. As stated in the introduction, our approach differs in the degree of integration of the various elements leading to an efficient easy to use prototyping system. The host CAD employed, RoboCAD, facilitates customisation of pull down menus and the addition of 'user' functionality either in the form of command strings or the running of external programs. VMC-CAD (as we refer to the integrated design and machine control system, Ref. 10) uses this facility together with a suite of specially written programs produced by Axiomatic Technology Limited.

Simple programs are provided for interaction with the machine controller, for example to change the feed rate during cutting. The user simply clicks on the appropriate 'set-up' pull down menu and modifies the feed rate which appears in the dialogue box. Many commands are direct and bypass any sequential commands which may be in the controller or communications buffer. Other direct commands include: pause for work inspection or tool change, cancel to flush the current job and depth controls.

Initiating machining of a component is as simple as starting a plotter, with 'machine plot' options being accessible within the 'output' pull-down. The innovation here is the way in which the PPD preprocessing software is automatically invoked. It is obviously desirable that contours should be cut in single operations regardless of what editing may have taken place. Equally, where cut material is to be cut through, it is important for inner contours to be cut before outer contours and for contours to be cut in an appropriate direction to avoid a poor quality edge on the cut component. All this and much more is automatically performed in a virtually invisible manner by the PPD software.

For some operations, however, particularly when machining complex components a degree of interactive control of the post processing may be required. In its interactive mode, the post processing software allows all features to be switchable and gives the user complete control over the machining process. This enables the specification of contour start points, the placement of additional/alternative holding tabs, specification of multi-layer and multidepth cutting (for thicker materials and three dimensional operations).

7. ACHIEVEMENTS

The system described has been used successfully to produce a range of parts for both small and very large model aircraft, including wing ribs for full-sized light aircraft. Normally, the production of these components would have been a highly skilled time consuming manual operation. Traditional CNC or CADCAM processes would not provide an alternative due to high costs and programming/skills overheads. The system produced enables the creative designer to rapidly verify design validity through production of accurately machined parts which can be used to build prototype models of all kinds.

The finished component - part of the parapet of a model bridge for a film set.

Following successful design validation, the system is also suited to batch production and indeed several well known model kits for larger aircraft incorporate parts cut on such a machine system.

Most discussion in this paper relates to the production of aircraft components, such as wing ribs and fuselage formers. However, the capability of the machine is not limited and the system has been successfully used to machine three dimensional mould plugs from solid material. The examples illustrated naturally involved significant design time, but the result in the form of the fabricated bridge used in production of a recent movie more than justified the investment.

8. FURTHER DEVELOPMENTS

Having proved the functionality of a cartesian robot in some areas of prototype component manufacture, the authors have recognised some limitations of their system for three dimensional work. Whilst on economic grounds not wishing to move to 5 or more axis machines, they wish to further exploit the potential of the three axis machine. This will be achieved by the implementation of more comprehensive three dimensional cutting motions and the incorporation of an inexpensive automatic tool change mechanism.

Since there are obvious advantages in a prototyping machine which is also commercially suited to production work, features to further increase productivity are envisaged. Amongst these are multiple spindles, each cutting the same pattern from a number of sheets of material. Further improvements are possible by cutting several layers of material at once, which is possible using the vacuum clamping system and the tabbing facility to keep the workpiece stable in the X-Y plane. Furthermore, a Z-plane pressure damper bearing on the workpiece around the cutting head will stabilise the upper sheet against upward forces created by the swarf extraction system and the spiral of the cutting edge.

Whilst there are of course many operations which will still require the use of very sophisticated machinery, the system will satisfy the requirements of a wide class of applications at an economic cost not approachable by alternative techniques. Other related work being undertaken by the authors includes direct vision based machine control for a different class of cutting operations. Here the use of pre- and post-process cameras combined with innovative neuro-fuzzy control (Refs. 4, 5 & 6) adds to the potential.

9. CONCLUSIONS

The work presented in this paper demonstrates a practical approach to the realisation of 'desktop manufacture'. The ease with which graphic design can be translated to machined 2 and 3 dimensional components, with examples from the production of miniature aircraft, indicates the validity of the approach. The versatility of a cartesian robot cutting machine, combined with powerful yet easy and quick to use design facility results in a competitive approach to the production of prototype and batch production components.

Due to the accuracy of manufacture, parts with complicated sections can be built up from smaller parts, which accurately lock together to form the final assembly. This can be used in conjunction with traditional techniques to get the best of both worlds, particularly with large sub-assemblies.

The accuracy and consistency of *this* machining system means that rapidly produced prototypes from it are consistent with production components from the most sophisticated production processes, particularly as the correct material can be used, giving a truly working prototype.

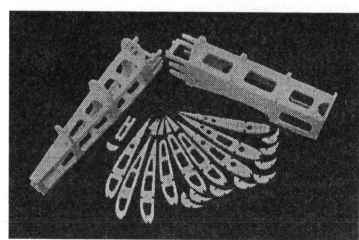

A set of wing ribs and fuselage components cut from card-backed foam, and partially assembled. Other materials typically used are balsa, plastics, aluminium and brass.

10. Acknowledgements

The authors gratefully acknowledge the support of EPSRC and HEFC. The support of Pacer Systems Limited, manufacturers of the cutting machine and Axiomatic Technology Limited, developers of the specialist software is appreciated. Much technical input came from Sequoia Systems Limited.

References:

1 "An Improved Method for Automatic Fairing of Digitally Represented Two Dimensional Cutter Paths"
Proc of 1994 Engineering Systems Design and Analysis Conference, Vol 5, pp 199-204, presented at 2nd Biennial European Joint Conference on Engineering Systems Design & Analysis, ESDA, London, England - July 4-7 1994
J F Poliakoff, P D Thomas

2 "An Improved Algorithm for Automatic Fairing of Non-Uniform Parametric Cubic Splines"
Accepted by Computer Aided Design (to be published)
J F Poliakoff

3 "Pacer Systems Ltd. The Growth of a Company - The Development of a Department"
Proceedings of Technology Transfer & Innovation Conference, 18th July 1994, Queen Elizabeth II Conference Centre, London
P D Thomas, N Sherkat

4 "Real Time Vision for Lace Cutting"
Proc of Machine Vision Applications in Industrial Inspection II, International Symposium on Electronic Imaging: Science and Technology, IS7T/SPIE 94 Vol 2183, pp 322-333, San Jose, CA, USA, 6-10 Feb 1994
N Sherkat, M Birch, P D Thomas

5 "A Fuzzy Reasoning rule-based system for handling lace pattern distortion"
Proc of Machine Vision Applications in Industrial Inspection III, published by SPIE, Vol 2423, pp323-333, 1995
N Sherkat, Chi-Hsien V Shih, P D Thomas

6 "A fuzzy reasoning rule-based system for handling lace pattern distortion"
IS & T/SPIE Symposium on Electronic Imaging: Science and Technology, San Jose Convention Centre, USA, 5-10 February 1995
N Sherkat, Chi-Hsien V Shih, P D Thomas

7 "A Stepping Motor Control algorithm for smooth continuous path motion"
IEE International Conference, Control '94 - March 21-24 1994, University of Warwick
W Steiger, N Sherkat, PD Thomas

8 "Smooth Continuous Path Motion through Variable Pulse Control"
Proc of ASME 2nd Biennial European Joint Conference on Engineering Systems Design & Analysis (ESDA), Vol 64-6, pp 109-116, London, England, 4-7 July 1994
W Steiger, N Sherkat, P D Thomas

9 "Smooth Continuous Path Motion Generation for Stepping Motors",
3rd IFAC/IFIP Workshop on Algorithms and Architectures for Real Time Control, Ostend 31st May-1st June 1995, (to be published)
W Steiger, N Sherkat, P D Thomas

10 "VMC-CAD User Manual",
Axiomatic Technology Ltd , 1995,
D J Moore

welding fabrication: strategies for sloped surfaces

NORMAN and **P M DICKENS**
University of Nottingham, UK

SYNOPSIS

A 3D welding system at the Manufacturing Engineering and Operations Management department at Nottingham University is described. CAD designed objects are made out of weld beads built up on top of eachother, layer by layer, until the desired shape is produced. The path that the bead must be deposited on is planned using custom designed software. One particular problem is the building of sloping surfaces so that the molten welding material does not flow off its desired position before it has solidified. Experimental results of building sloping surfaces are presented and possible solutions discussed.

INTRODUCTION

Description of the 3D welding system at Nottingham University.

The 3D welding system at Nottingham (see photo. 1 and diag. 1) is part of a BRITE EURAM project (see acknowledgement) to research and develop methods of rapidly producing prototype designs by metal deposition.

CAD models (see photo. 2) are produced using Intergraph EMS software [1] on a Sun Sparc 10 workstation. These are sliced and the path for weld beads planned (see diag. 2) using extensions to Intergraph software [2].

The path coordinates, together with welding parameters picked from the welding database for the chosen bead dimensions, are passed to GRASP robot simulation software [3],[4] using a neutral STEP file format. The welding database consists of a list of welding parameters for each bead with its set of dimensions. The welding parameters are settings passed to the welding power supply, such as wire feed speed and pulse frequency, which control the physical condition and dimensions of the weld bead.

GRASP simulates the movement of the robot and the deposition of the weld bead (see photo. 3) to test the feasibility of the build movements and produces a program in the robot language for the Siemens RCM robot controller. The robot is a large six axes gantry robot with a remote two axes tilt and rotate table [5]. It is equipped with a Binzel welding torch attached to a welding wire feeder and a Fronius programmable power supply [6].

Pulsed mode MIG welding [7] is the type of metal deposition method used. This has proved to be the most suitable because of the lower heat build-up and the ability to control all important parameters.

Monitoring the welding process is very important. Voltage, current and wirefeed speed sensors are used. The incoming data is electronically conditioned and fed into a data acquisition card in a Pentium P66 computer. Labview software from National Instruments [8] displays, analyses and records the data.

PATH PLANNING AND SLICING

Slicing

Slicing is a process in which a CAD object (diag. 2a) is mathematically sliced horizontally from its base to the top in identical increments of height (diag. 2b). The CAD object is sliced using an incremental height selected manually from the GUI (Graphical User Interface) menu. (see photo. 4) bearing in mind the possible bead dimensions. The most important bead dimensions are height and width. (see diag 3). These are derived from cross-sectional images (diag. 4). This is very similar to the stereolithographic process [9], except for the much larger step heights and the direct conversion of the CAD object into sliced sections in 3D welding.

Path Planning

After the slicing process is complete the path for the welding torch (i.e. for the weld bead to be deposited) to follow must be planned (diag. 2c). The path planned corresponds to the position of the tool centre point (TCP) which is defined as being 10 mm from the end of the welding torch. This is accomplished by software algorithms using the selected strategy which is determined by the needs of the welding process. A simplified CAD object (photo 5) is shown sliced on one level and path planned (photo 6).

Accurate placement of the weld bead depends on the stability of the welding arc. The arc consists of a highly energetic ionised plasma which is entrained by the local electromagnetic field lines. Assymetrical placement of metal about the vertical axis of the welding torch will result in assymetrical field lines which distort the arc chaotically. Path planning strategies are based on, wherever possible, symmetrical deposition of material, so that the arc is stable and well behaved. In photograph 4 the preferred method of path generation is shown selected; alternate beads seperated by a gap are deposited and then an infill bead placed in the gap (diag. 3).

A complex object has straight or curved vertical sloping sides, as opposed to a simple object which has no sloping sides.

STATEMENT OF PROBLEM

In the case of sloping vertical walls, if the angle of the torch axis is maintained normal to the baseplate, the deposition symmetry is not preserved. The bead may well be displaced and molten metal from the weld pool may fall off (diag 5a). To avoid this it is necessary to build beads vertically on top of eachother (diag 5b). To do this the robot's tilt and rotate table is employed to orient the object being built so that this is achieved. This method, because of the slope angle (θ), means that the vertical incremental height is reduced by a factor of $\cos \theta$. Therefore some strategy to maintain the expected vertical height is needed.

Stereolithographic processes [8] cannot tilt the fluid bed, and so support structures must be built. With the ability to tilt and rotate the baseplate, this is not necessarily so with 3D welded objects.

Experimental procedure

The robot was programmed to produce vertical walls at various angles of slope (diag. 6). Also $0°$ slope walls were built at single bead incremental stages (diag. 7).

Mild steel was used and a proven set of welding parameters was chosen to produce a reliable bead.

The welded parts were sectioned, polished, and scanned at a resolution of 600 dots per inch. These images were directly measured and the results analysed.

Experimental results

Diagram 6 shows scanned images of walls built at various tilt angles. Diagram 7 shows vertically built walls from 6 to 12 beads high (the first few beads tend to be narrower and so are ignored, since they cool more quickly by conduction because of the shorter thermal path to the baseplate, which is in general at a much lower temperature than the newly deposited bead, the dimensions of the bead come to equilibrium after about 10 mm of build height). These images are superimposed on eachother to show the shape of the incremental bead at each level (diag. 8).

The measured bead heights were plotted against expected values (table 1). Notice that the measurements were taken at the TCP position which is defined as shown in diagram 9. With sloping walls the actual highest point is higher than the position of the TCP, but this is unimportant.

Table 1: Vertical build heights at various angles of slope.

θ	Theoretical height (mm) (22.4 × cos θ)	Experimental height (mm)
0	22.4	22.4
10	22.05	21.5
20	21.05	21.0
30	19.39	19.0
45	15.84	15.0

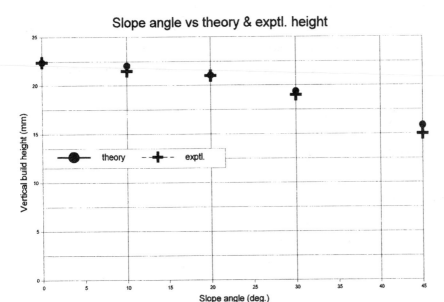

Slope angle vs theory & exptl. height

Interpretation of experimental results

The measurements from the scanned images show that the expected results are satisfactory within the limits of experimental error.

Conclusions from experimental results

The tilt and rotate table is demonstrated to enable the build of sloping walls in a predictable way.

PROPOSED SOLUTIONS

In order to maintain the vertical build height as described earlier two methods are suggested.

1. Select bead with greater height

A weld bead of greater height (multiplier of $\cos^{-1}\theta$) is selected from the welding database. Only the edge bead need be so. The infill and pitch beads can remain with the incremental slicing distance as their height dimension.

2. Same bead but use different slicing increment

Labein's slicing software must have incorporated the ability to alter the incremental slicing height automatically dependent on the slope angle. This is only possible with hollow objects,because with solid objects different infill beads with smaller heights are needed, this is more complex than solution 1. Major changes to the software algorithms would be required.

Solution 1 is the preferred option.

CONSEQUENCES & CONCLUSIONS

The slope angle (θ) must be determined automatically,for each slicing level, at the path planning stage.This needs to be done for edge beads only. Changes in the path planning algorithms need to be incorporated before this can be achieved. In the meantime manual alterations to the robot programs will be made to prove that these conclusions are correct.

References

[1]   (EMS Intergraph manual)

[2]   (Labein software manual)

[3]   (GRASP manual)

[4]   J. Norrish; Chapt.11, section 11.9.1; Advanced Welding Processes; pub. Inst. of Physics 1992; ISBN 0-85274-325-4

[5]   (robot operating manual)

[6]   (Fronius power supply manual)

[7]   J. Norrish; Chapt. 7; Advanced Welding Processes; pub. Inst. of Physics 1992; ISBN 0-85274-325-4

[8]   (Labview for Windows manual)

[9]   Paul F. Jacobs; Rapid Prototyping and Manufacturing-Fundamentals of Stereolithography; pub.Society of Manufacturing Engineers; ISBN 0-87263-425-6.

Diagrams

(1)   Schematic of robot welding system

(2)   CAD object - unsliced, sliced, and one layer path planned

(3)   Bead dimensions illustrated   (a) base bead

(b) incremental bead

(c) infill bead

(4)   Scanned images of beads-   base beads

-    multiple beads

(5a)   Sloping wall built with horizontal baseplate

(5b)   Sloping wall built with tilted baseplate to prevent bead collapse

(6)   Scanned images of beads built at various tilt angles

Photographs

(1)   General view of robot welding cell.

(2)   Connecting rod CAD object.

(3)   GRASP image of B&C robot.

(4)   Labein slicing and path planning GUI.

(5)   Simplified CAD object.

(6)   CAD object sliced and path planned.

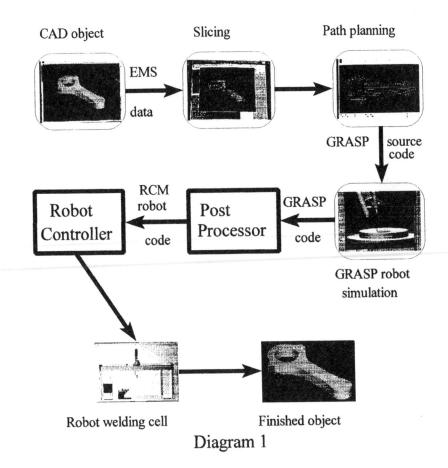

Diagram 1

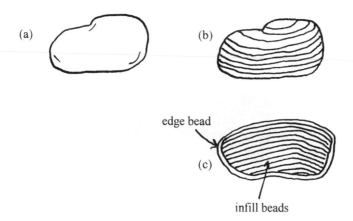

(a)

(b)

edge bead

(c)

infill beads

Diagram 2   CAD object

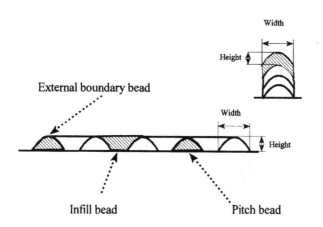

Width

Height

External boundary bead

Width

Height

Infill bead

Pitch bead

Diagram 3

Description of weld bead dimensions
and nomenclature

Diagram 4: Scanned single weld bead

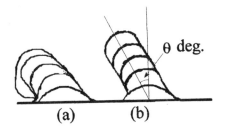

(a)　　　　　(b)

θ deg.

Horiz. baseplate.　　　Baseplate at θ deg.

Diagram 5

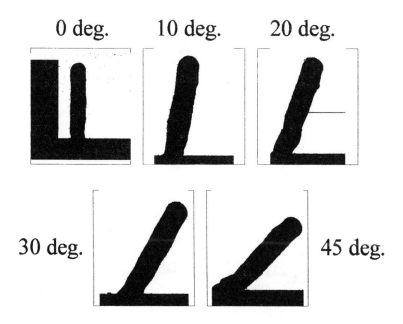

0 deg.   10 deg.   20 deg.

30 deg.   45 deg.

Diagram 6: Sloping beads

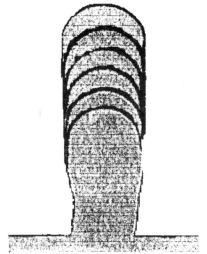

Diagram 7: Superimposed beads

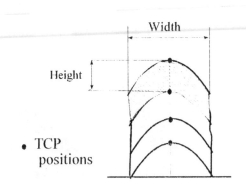

Diagram 8

Photograph 1

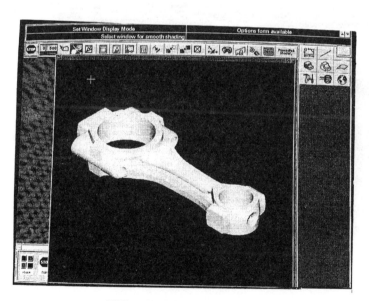

Photograph 2

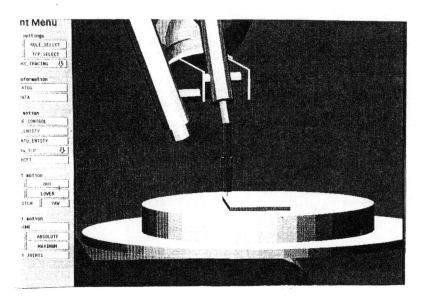

Photograph 3

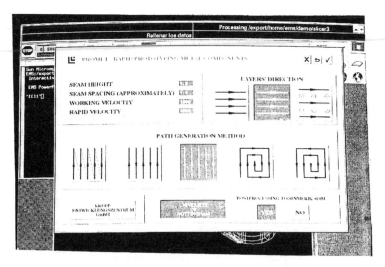

Photograph 4

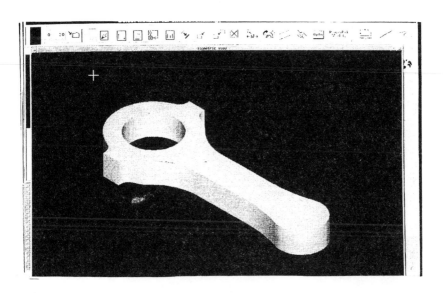

Photograph 5

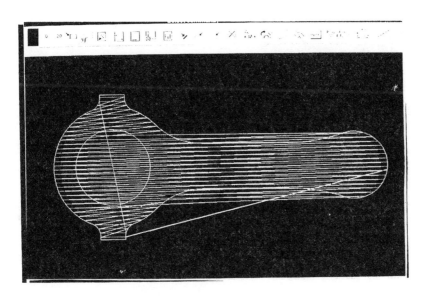

Photograph 6

The development and application of a stereolithography build simulator

JELLEY and C P THOMPSON
Cranfield University, UK

Abstract: A number of barriers to the widespread adoption and application of stereolithography by companies are discussed. These must be addressed if the technology is to progress from being a piece of specialised machinery and gain universal acceptance. The development of a simulator is seen by the authors as an important step towards this. An accurate, usable simulation will ensure a better understanding of the physical and chemical processes involved in stereolithography. It will also allow users and potential users the opportunity to test new parameters and build styles and to evaluate model characteristics prior to an actual build. The early development of the simulator is discussed. At present it incorporates a two-dimensional computational model of a static laser-induced photopolymerization process. This process has many key features found in modern stereolithography apparatus.

The initial results presented show good agreement with theory and the other work in this area. Validation and future developments of this model are discussed.

1   Introduction

The last decade has seen the worldwide adoption of numerous additive fabrication[1] techniques by a large number of companies. These methods, which allow complex physical three-dimensional models to be generated from a computer representation, have been shown to aid in the reduction of product development lead times and costs, and the improvement of product quality. The last ten years has also seen the rapid development of these technologies and the demand for models with better accuracy and a greater functionality. This has led to the development of new processes and new materials, along with a greater understanding of the physical and chemical interactions involved with each process. Models made from these technologies today range from those used purely for visualisation purposes to those used for the production of master models for sand and investment casting. There are however, a number of issues arising from this rapid development that are currently acting as a barrier to the adoption by smaller companies, and the application by companies worldwide. These issues must be addressed if the technology is to evolve from being a piece of specialised machinery. Discounting the obvious economic issue of the total cost of purchasing one of these machines, we will briefly discuss five of the more 'technical' issues (focusing particularly on their implications for stereolithography)[1] and how the development of an accurate numerical simulation will address them. We will then briefly discuss the early research on stereolithography process modelling, and how the problems of modelling have so far been attempted. A process model will then be introduced which, based on some of this earlier research, will become the basis of a stereolithography simulation. Early results are presented and future developments discussed.

1.1 Current issues

The following section deals with five of the main issues arising from the rapid development of stereolithography and other Rapid Prototyping (RP) technologies over the last decade. Each of these issues has a bearing on the widespread adoption of this new technology by smaller companies and its application worldwide.

Complexity of process fundamentals. Research initiated in the early 1990s by 3D Systems Inc. led to the documentation of the fundamental relationships involved during the interaction of actinic photons with a reactive photopolymer[2] [3]. A thorough knowledge of the conclusions drawn in these papers is (at present) considered a crucial component in any attempt to build accurate, useful parts by stereolithography. However, most companies using or (perhaps more importantly) contemplating the use of stereolithography, lack the technical expertise required to interpret the fairly complex physical, mathematical and chemical theory contained in these conclusions, as well as the time and money needed to apply them by experimental model building. The nature of these fundamentals has led to the current situation among existing users where useful model building relies mostly on experience, or an unswerving allegiance to the default parameters. This tends to make model building (for the less experienced) a time consuming and expensive process of iteration, with parameters requiring subtle manipulation before and after each run, to obtain a model with the required characteristics [1] [4]. The complexity of these fundamentals, coupled with the time required to gain the valuable experience in model building, tends to dissuade many potential small users from investment. Most tend to prefer the services offered by the increasing number of RP bureaus, and are currently awaiting a technology developed to the stage where it can be used almost as a printer type peripheral[5][6].

Increasing number of available technologies. There are currently a large number of RP technologies available on a commercial basis. Each offering 'faster model build times', 'greater accuracy', 'improved repeatability' and 'increased model functionality'. Even companies offering machines with the same basic

---

[1]The term stereolithography will be used to describe the genre of methods which are based on the selective photopolymerization of successive layers using optical energy.

underlying technology, such as stereolithography, possess a bewildering range of process capabilities[7]. It has been claimed that this wide choice confuses the potential buyer and, as mistakes can be costly, acts as a barrier to the adoption of RP by smaller companies. Each vendor claims to meet customer requirements, while offering machines at a wide variety of prices and using a range of model building techniques.

Increasing choice of materials available.   Initially, stereolithography resins were marketed by Ciba-Geigy and DSM-Desotech. These resins, Cibatool XB5081 and Desolite 801, were patterned after traditional coating materials, containing an epoxy diacrylate as monomer and N-Vinyl Pyrrolidone (NVP) as a reactive dilutant. Whilst these resins served their purpose in the early years of stereolithography, they gradually proved deficient as the models they produced could only be used as conceptual design models. These deficiencies included: the brittleness of the cured polymer, high resin viscosities and high volume shrinkage. Research and development of photopolymers, initiated by Ciba-Geigy in 1988, brought about major changes in the characteristics of all stereolithography resins and by mid-1991 most of the early problems had been addressed. These new resins had greater impact resistance, enhanced mechanical properties and the ability to be machined. Greater understanding of both stereolithography and the build parameters affecting model characteristics led to increased industrial interest in multi-purpose models. New resins were therefore developed, which (when used with certain build parameters) produced models with characteristics relevant to the model's intended function. This increasing choice has allowed a wider user base to develop and offers existing users a number of benefits. However there are a number of disadvantages; for example, the large resin selection appears to offer the potential user similar model and build characteristics. For the existing user, each resin and its associated properties need to be carefully understood, in order to obtain the desired model characteristics. This often leads to numerous experimental builds of a well known model by an experienced user.

Complexity of model building process.   Many people new to RP still expect the transition from a 3D CAD model on the computer screen to a physical model to be as simple as 'pressing a button'. The reality is more complex. The model building process adopted by 3D Systems Inc. in their StereoLithography Apparatus (SLA) goes through roughly eleven separate stages each requiring a certain amount of human intervention before the next stage can commence.

The changing function of prototypes.   Increasing model accuracy brought about by process improvements and a wider choice of resin has led to the explosion of new applications for RP. We have categorised these into five main areas:

- Prototypes:— visualisation and verification.

- Test:— functional fit and assembly.

- Part:— real plastic components.

- Design:— iteration, optimisation.

- Tooling:— metallic and non-metallic from models.

The problems associated with the model's changing function within the stereolithography industry are mainly concerned with resin selection, parameter, accuracy and resolution issues. These issues need to be fully addressed prior to each build. Changes of resin require a change of build parameters and these parameters ultimately affect the model's accuracy and resolution, each of which are important in the intended function of the model. Model building again becomes an expensive process of iteration, this time for existing experienced users, as they try to make the process deliver the required model characteristics. This increasing functionality has caused a change in user profiles in recent years. RP, firmly rooted within mechanical engineering since its conception, has started to encroach upon areas such as medicine, fluid dynamics and even art. To exploit these areas fully the RP community must simplify the process of model production. At present no training method exists to enable new users to practice off the machine before physically building a model. Moreover, no RP method incorporates any form of intelligent process control, which removes the human from the decision-making process.

1.2   The need for a numerical simulation

The issues discussed mean that many existing users still only have a basic understanding of stereolithography and the fundamental parameters which affect model build. Many still have to make several runs to obtain acceptable results. They have to modify the manufacturer's recommended parameters based on their own experience, re-orientate the part, correct for shrinkage, interpret material chemistry quirks and re-set the laser control parameters. Many also have problems with machine repeatability; one model built directly after another with the same parameters does not guarantee uniformity. This process of continuous iteration can be expensive, delaying the production of other work and tying up machinery which needs a high throughput to be cost effective. Time must be set aside on the machine for training and experimentation as new technologies, new resins, process improvements and new uses for the models emerge.

New users are in much the same position. As qualified people (not only familiar with CAD/CAM modelling but also the whole process) have been identified as making the difference between a successful operation and a disaster [6], the need for training and education is vitally important. Currently there is no substitute for hands-on experience

2 The research project

The aim of this project is to develop a software package that will allow a quantitative analysis of the physical and chemical processes in stereolithography to be performed. Paying particular attention to the recent research into the numerical modelling of the curing process, it will help to address some of the issues outlined. The software developed will provide:

- A research tool. Incorporating a detailed mathematical model of the curing process it allows:

  - A better understanding of the curing process — macro-level information on strength, surface finish and accuracy will be obtained.

  - Model Evaluation — the simulator will allow an evaluation of parameters prior, during and after the build, giving a prediction of the model's accuracy as well as its mechanical properties. Probable deviations of the actual model from the ideal can be assessed by accurate simulation before time and money is spent physically building.

  - Build Evaluation — will provide an estimation of build times, cure profiles, distortion and shrinkage, as well as giving an indication of the temperatures throughout the vat as polymerization progresses and the progression of cure with time.

  - Technology evaluation — when incorporated with other process models this will allow future users to decide which technology to adopt for each individual application.

  - Sensitivity analysis — unlike physical builds, simulations can be performed in a matter of minutes, individual parameter values can be precisely changed and the results investigated, allowing users to optimise the process.

- A platform for the testing of new resins, build styles and process improvements.

- The development of a standard set of build parameters and resins for some predetermined model characteristics.

- A means to train new SL users off the machine in the effects of process parameters on final model characteristics.

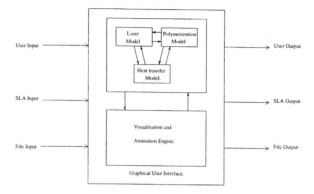

Figure 1: Basic numerical simulation framework.

This paper will continue with a discussion on the work done to date on this project, in particular the development of the numerical model of the curing process which will become the basis of the simulator (**Figure 1**).

3 Background

At the beginning of the research project, we examined the current trends in Rapid Prototyping and in particular the work on modelling the physical and chemical processes in stereolithography. We concluded from these studies that modelling could be split into roughly three main areas:

1. modelling of the laser;

2. modelling of the photo-initiated free radical polymerization;

3. modelling of the heat transfer involved in the process.

3.1 Modelling of the laser.

Modelling of the laser beam and how its intensity decreases with depth in a vat of resin has over the last five years been the subject of a number of research projects[8][9]. For instance, in their paper Brulle et al. used the Beer-Lambert Law to suggest an ideal model of laser-induced polymerization, going on to discuss some probable causes for the deviation from this ideal case (**Figure 2** [8]). A Monte Carlo simulation model was then developed to test the hypothesis stated. These declared that the occurrence of a laser beam with an average Gaussian flux density, but a random space-time evolution, coupled with a polymerizable material which is partially heterogeneous, may lead to the phenomena witnessed

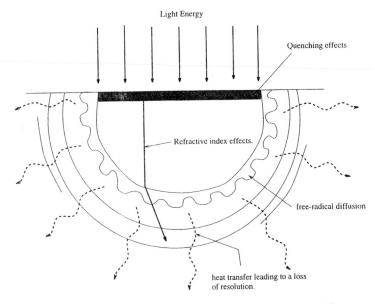

Figure 2: Phenomena associated with photo-reaction[8].

experimentally. The initial model takes into account the change of refractive index as the liquid monomer becomes solid polymer and the possible photobleaching effect. Their results seem to follow the experimental observations and describe a cure profile which is 'torch-like' rather than parabolic. Other work[10], suggests a mathematical representation based purely on the optical analysis of the laser beam path as it travels through the photosensitive resin, calculating the exposure distribution in the resin. Using this model, a solidified profile is determined by a threshold level of exposure.

3.2 Modelling of photopolymerization.

Modelling of the photopolymerization inherent within stereolithography has its roots firmly in the polymer science of radiation curing. The basic kinetic models proposed by Decker[11][12][13], Fouassier[14][15][16] and Odian[17] use a set of equations which describe the three steps that make up the laser induced radical chain polymerization process i.e. initiation, propagation and termination (**Figure 3**).

Models based on these equations[18] give an accurate picture of the polymerization process as it progresses over time; from these equations, values for the rate of initiation and rate of polymerization can be achieved. Early results using these models suggest that they reasonably describe the photochemical processes involved in stereolithography. They show a relatively good correlation with theory and demonstrate (amongst other observations):

- That the percentage conversion of monomer to polymer varies as a function of position and time.

- That this conversion reaches a maximum of approximately 60% as reaction rate becomes diffusion controlled.

- That polymerization produces 'bullet shaped' cured profiles.

The problems with these models are:

- The models are only an approximation to the real situation and incorporate some assumptions.

- The theory underlying these models is still a matter of much research.

- The equations do not directly relate to stereolithography and the user is expected to have a certain amount of knowledge of polymer science in order to understand them.

3.3 Modelling of heat transfer.

Two strands of research into the modelling of the heat transfer within the process have been discussed in recent years. One is concerned exclusively with the conduction of heat by the resin due to the exothermic polymerization process[19], while the other is also concerned with the modelling of a laser light source which adds heat to the resin, such as a thermal IR laser[20]. Both have been approached in a number of similar ways, each requiring a knowledge of:

- The target's optical and thermal properties during the irradiation.

- The laser beam distribution.

- The dynamics of the irradiation process.

- The processes of phase change in the target material.

Both assume that the material is homogeneous and isotropic and that the thermal properties of the resin do not vary greatly with temperature and are therefore assigned an average value for the temperature range studied. Other heat transfer effects such as forced and free convection have so far been neglected.

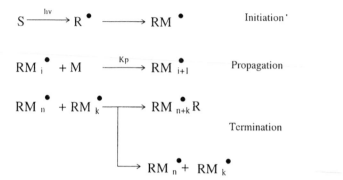

Figure 3: Classical steps in photo-initiated free-radical polymerization[14].

4   Current Model

The current model we are using as the initial basis of the simulation is based upon the two-dimensional, static laser model proposed by Flach et al. For reasons of space we have restricted our description of the mathematical model. Further details can be found in the following references[18][21][5][22]. The equations which make up this model describe the physical and chemical changes that occur in a small cylindrical volume of monomer mixture illuminated by a stationary laser (**Figure 4**). These equations are solved numerically within a computer program which allows the selection, by the user, of a number of parameters which affect the process. The outputs of the program include spatial and temporal variations in the concentration of monomer and polymer, depletion of the photoinitiator, and the local variations of temperature in and around the region contacted by the laser light[19].

Since the model describes a real physical process a number of assumptions had to be made to simplify its development. These are (see [19][2][8]):

- The photopolymer resin obeys the Beer-Lambert law of exponential absorption.

- The laser irradiance distribution is Gaussian.

- The flow of material due to convection or diffusion in any direction is negligible.

- There are no light scattering, diffraction, refraction or reflection effects.

- The chemical reaction occurs only in the cylindrical region of radius R and depth D.

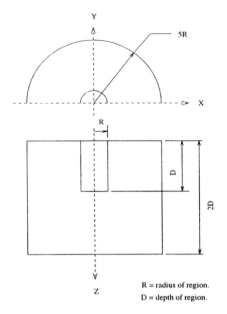

Figure 4: Coordinate system used in initial model.

- The conduction of heat occurs over a cylindrical region of radius 5R and depth 2D.

- The heat generated is due to heat of polymerization only.

- The heat loss from the surface of the vat is negligible.

- All physical and thermodynamic properties are independent of degree of conversion and temperature (i.e. constant).

- A square root dependence of the rate of polymerization on the light intensity is assumed to continue throughout polymerization. This stems from the assumption of steady-state production of all radical species and that the termination step is purely bi-molecular.

- The evolution of the polymerization zone is unaffected by volume shrinkage.

- The evolution of polymerization is unaffected by oxygen inhibition.

The model consists of four partial differential equations (PDEs), one governing the rate of change of light intensity with respect to depth (equation 1) two of which govern the radical-chain polymerization which occurs in the region (equations 2 and 3), and one of which governs the heat transfer in the resin due to the exothermic reaction (equation 4). The incident laser light is assumed to have a Gaussian intensity distribution:

$$I(x, y, z, t) = I_o e^{-2\left(\frac{r}{w_o}\right)^2} \text{ at } z = 0,\ r \geq 0,\ t \geq 0$$

where:
$I(x,y,z,t)$ = light intensity ($E\ cm^{-2}\ s^{-1}$)
I_o = peak light intensity ($E\ cm^{-2}\ s^{-1}$)
r = distance from centre of laser beam (cm)
w_o = beam radius (cm)
z = depth from the top of the region (cm)
t = time from when laser is on. (s)

The decrease in light intensity with depth is given by:

$$\frac{\partial I}{\partial z} = -\varepsilon S I \tag{1}$$

where:
$S(x,y,z,t)$ = photoinitiator concentration ($mol\ l^{-1}$)
ε = absorptivity of the photoinitiator ($l\ mol^{-1}\ cm^{-1}$)

The light absorbed by the photoinitiator is given by:

$$I_a(x, y, z, t) = \varepsilon S I$$

where:
$I_a(x,y,z,t)$ = absorbed light ($E\ cm^{-3}\ s^{-1}$)

The rate of initiation is given by:

$$r_i = \phi I_a$$

where:
ϕ is the quantum yield for photoinitiation.

The rate of change of the photoinitiator concentration is given as:

$$\frac{\partial S}{\partial t} = -\phi I_a \tag{2}$$

The rate of change of monomer concentration can be calculated from:

$$\frac{\partial M}{\partial t} = -R_p$$

where:
$M(x,y,z,t)$ = monomer concentration (mol l^{-1})
R_p = polymerization rate (mol l^{-1} s^{-1})
Rate of polymerization is given by:

$$R_p = k_p M \left[\frac{\phi I_a}{k_t} \right]^{0.5}$$

where:
k_p = propagation rate constant (l mol^{-1} s^{-1})
k_t = termination rate constant (l mol^{-1} s^{-1})
The rate of change of monomer concentration finally becomes:

$$\frac{\partial M}{\partial t} = -k_p M \left[\frac{\phi I_a}{k_t} \right]^{0.5} \tag{3}$$

The temperature field in the vicinity of the exposed region is given by a transient heat conduction equation with a heat source term:

$$\rho C_p \frac{\partial T}{\partial t} = \kappa \left[\frac{1}{r} \frac{\partial}{\partial r} \left(r \frac{\partial T}{\partial r} \right) + \frac{\partial^2 T}{\partial z^2} \right] + \Delta H_p R_p \tag{4}$$

with:

$$
\begin{aligned}
T(r,z,t) &= T_o \text{ at } t = 0,\ z \geq 0,\ r \geq 0 \\
\frac{\partial T}{\partial r} &= 0 \text{ at } r = 0,\ z \geq 0,\ t \geq 0 \\
T(r,z,t) &= T_o \text{ at } r \to \infty,\ z \geq 0,\ t \geq 0 \\
T(r,z,t) &= T_o \text{ at } z \to \infty,\ r \geq 0,\ t \geq 0 \\
\frac{\partial T}{\partial z} &= 0 \text{ at } z = 0,\ r \geq 0,\ t \geq 0
\end{aligned}
$$

where:
ρ = Density (g cm^{-3})
C_p = Heat Capacity (J g^{-1} K^{-1})
k = Thermal Conductivity (W cm^{-1} K^{-1})
ΔH_p = Heat of polymerization (J mol^{-1})

5   Results

Early results from the model are in relatively good agreement with Flach et al. and show a good correlation with the theory[17][11][12][13][14].

Figure 5. shows a graph of degree of conversion with respect to time. Two

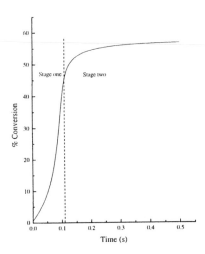

Figure 5: Graph showing degree of conversion with respect to time.

stages of the polymerization process can clearly be seen. In stage one the chain reaction develops rapidly up to about 45% conversion. This is followed by a progressive slowing down (stage 2) which is attributed to the increasing viscosity of the polymer system reducing the mobility of the reactive sites and the decreasing probability of an encounter between the polymer radicals with the reactive double bonds. Actual polymerization profiles recorded by real time infra-red (RTIR) spectroscopy would produce similar results but would have an initial induction period, due to the presence of atmospheric oxygen which is known to strongly inhibit the polymerization by reacting with the initiating radicals as well as with the growing polymer radicals [12].

Figures 6 and **7** show the concentration of photoinitiator with respect to time. **Figure 6** shows the change in photoinitiator concentration at the axis of the region (r = 0) with varying depth and **Figure 7** shows radially the change in photoinitiator concentration on the surface of the resin (z = 0). Both graphs clearly show the steady decrease of photoinitiator concentration as the polymerization progresses; this is in agreement with the theory. The photoinitiator plays a key role in light-induced polymerization in that it produces the free radicals with time which, by reacting with the monomer, initiate the chain process.

Figures 8 and **9** show how the monomer concentration changes with depth and radial position. The graphs show that, as expected, monomer concentration decreases with time throughout the region, until it reaches a minimum value of 1.30 mol l $^{-1}$. This value corresponds with the values obtained for percentage conversion.

Figures 10, 11, 12, 13 and 14 are contour plots showing the temperature distribution throughout the region over the time step calculated. Preliminary results are promising with temperature increasing rapidly due to the exothermic reaction to a maximum temperature of 68°C and heat being dissipated slowly throughout the region. Temperature rises to a maximum at the layer surface but decreases rapidly with depth, because polymerization is incomplete and inhomogeneous, since the light is, in principle exponentially absorbed. The results also show that heat transfer is greatest in the radial direction; this phenomena is unexpected and is currently under investigation.

Figures 15, 16, 17 and 18 are contour plots showing the percentage of monomer converted to polymer over the entire region. These show conversion reaching a maximum of 60%. If we assume that gelation occurs at approximately 50% conversion we see that the centre of the irradiated area requires less than 0.3 sec to be polymerized; plotting the contour at this conversion gives an indication of the cured profile, which as expected, has a parabolic shape. The maximum conversion reached also agrees with theory which states that the radical chain polymerization is never complete; there is always some unreacted monomer inside the polymer.

6 Discussions

The model presented in this paper contains the key elements of the photochemistry and thermal variations which are required to understand the stereolithography process. The results are not yet fully validated but they are in qualitative agreement with polymer theory and the work by Flach et al. Moreover, points of quantitative agreement have been identified, which is encouraging.

7 Conclusions

There are a number of outstanding issues which inhibit the widespread adoption of RP technology. These are related to the accuracy, repeatability and ease of use of the various processes. A numerical simulation will help to resolve several of these issues and we have presented a model of the curing process which forms the basis of a full simulator of an SLA. Preliminary results are in reasonable agreement with theory and experimental results, but some further investigation and development is needed to enable the model to be a useful tool for stereolithography research.

8 Future Work

The main thrust of this initial work was to develop the basis and framework of a stereolithography simulation and to briefly discuss the initial model which, when developed, will become the basis of this simulation. Future developments are planned which will improve the accuracy of the simulation results and improve the usability of the simulator. This will include:

- Incorporating the effects of Refractive index changes on cured profiles.

- A Thermal IR laser model.

- A non-vertical laser beam model to accomodate angle of incidence.

- An animation capability. Results from the simulation need to be interpreted rapidly, and with relative ease. The addition of an animation capability will allow the user to view and analyse results.

- A non-Gaussian laser model. Nearly all current models make an assumption that the laser beam intensity distribution is Gaussian. In reality this is far from the case, as a pure Gaussian mode implies the use of expensive control apparatus.

- Cure profile determination[19]. The observation of cure profiles is one of the easiest ways to validate the simulation results, real cure depths and line-widths can be measured against modelled. The observation of cure profiles will include not only single strings but also cured layers.

- A moving laser light source[19]. The development of the model will allow either a two-dimensional stationary laser light source or a three-dimensional moving one to be used as the basis for the simulation.

- The determination of resin working curves[19].

- A dark reaction model [19].

- Determination of temperature effects on model build.

9 Acknowledgements

The authors gratefully acknowledge the help of Lawrence Flach and Richard Chartoff at the University of Dayton Research Institute.

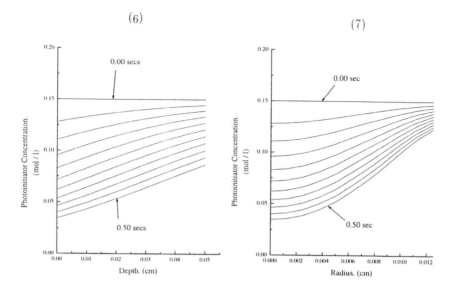

Figures 6 and 7: Showing the concentration of photoinitiator with respect to time, at r = 0 (Figure 6) and at z = 0 (Figure 7) at equal time intervals.

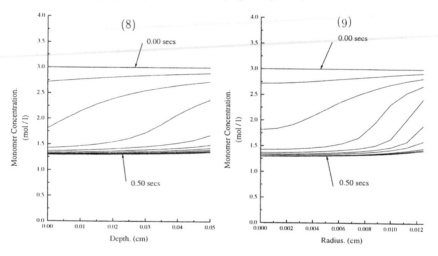

Figures 8 and 9: Showing the concentration of monomer with respect to time, at r = 0 (Figure 8) and at z = 0 (Figure 9) at equal time intervals.

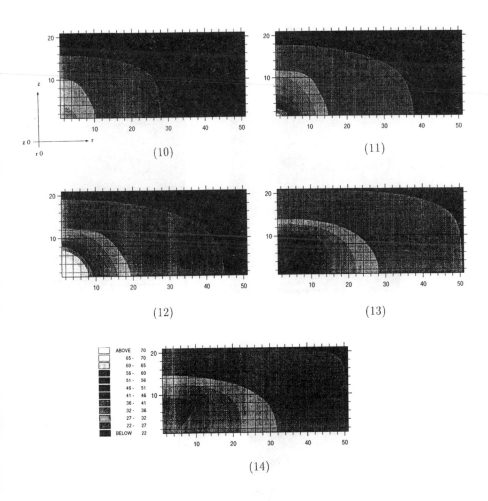

Figures 10, 11, 12, 13 and 14: Contour Plots showing temperature throughout the region, at t = 0.1 s, t = 0.15 s, t = 0.18 s, t = 0.30 s and t = 0.45 s.

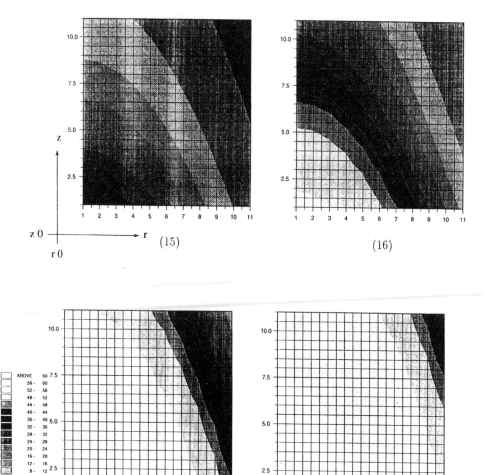

Figures 15, 16, 17 and 18: Contour plots showing percentage monomer conversion throughout the region at t = 0.1 s, t = 0.15 s, t = 0.30 s and t = 0.45 s.

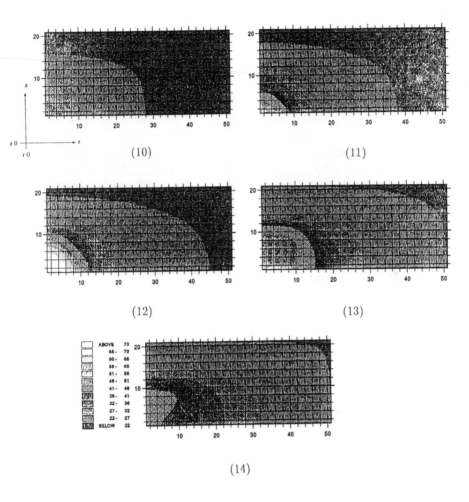

Figures 10, 11, 12, 13 and 14: Contour Plots showing temperature throughout the region, at t = 0.1 s, t = 0.15 s, t = 0.18 s, t = 0.30 s and t = 0.45 s.

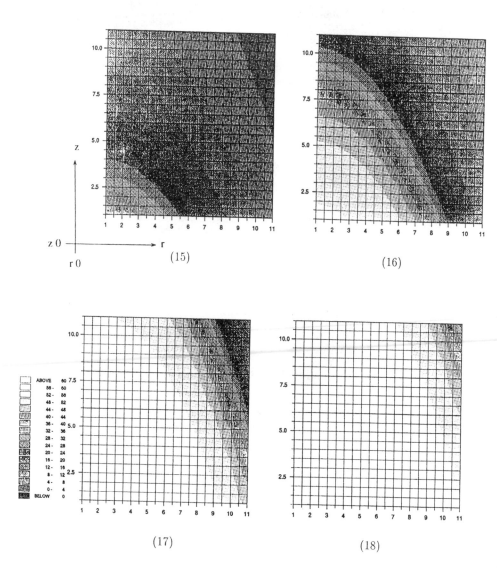

Figures 15, 16, 17 and 18: Contour plots showing percentage monomer conversion throughout the region at t = 0.1 s, t = 0.15 s, t = 0.30 s and t = 0.45 s.

References

[1] M. Burns. *Automated fabrication:Improving productivity in manufacturing.* PTR Prentice Hall, 1993.

[2] P.F. Jacobs. Fundamentals of stereolithography. In Dr.P.M.Dickens, editor, *Proceedings of the First European Conference on Rapid Prototyping.*, pages 1–17. University of Nottingham., July 1992.

[3] P.F. Jacobs. *Rapid Prototyping and Manufacturing:Fundamentals of Stereolithography.* Society of Manufacturing Engineers., 1992.

[4] B.E. Hirsch and H. Muller. Stereolithography - fields of application and factors influencing the accuracy. In D. Kochan, editor, *Complex Machining and AI-Methods*, pages 219–233. Elsevier Science, 1991.

[5] L. Flach and R.P. Chartoff. Stereolithography process modelling-a step towards intelligent process control. In *Proceedings of the Third International Conference on Rapid Prototyping.*, pages 141–147. University of Dayton., 1992.

[6] T. Wohlers. Rapid prototyping systems:buy now...next year...or never? In *Proceedings of the Third International Conference on Rapid Prototyping.*, pages 193–198. University of Dayton., 1992.

[7] G. Lart. A comparison of rapid prototyping systems. In Dr.P.M.Dickens, editor, *Proceedings of the First European Conference on Rapid Prototyping.*, pages 243–254. University of Nottingham., July 1992.

[8] B. Brulle Y. Bouchy, A. Valance and J.C. Andre. Industrial photochemistry xxi. chemical, transport and refractive index effects in space resolved laser photopolymereization. *Journal of Photochemistry and Photobiology A: Chemistry.*, 83:29–37, 1994.

[9] H. Narahara and K. Saito. Characterization of solidified resin created by three-dimensional photofabrication. In *Proceedings of the Fourth International Conference on Rapid Prototyping.*, pages 271–282. University of Dayton., June 1993.

[10] H. Narahara and K. Saito. Fundamental analysis of a single layer created by three-dimensional photofabrication. In *Proceedings of the Third International Conference on Rapid Prototyping.*, pages 113–120. University of Dayton., July 1992.

[11] C. Decker and K. Moussa. Kinetic investigation of photopolymerizations induced by laser beams. *Makromolekulare Chemie.*, 191:963–979, 1990.

[12] C. Decker. Laser induced polymerization. In *Materials for Microlithography.* 1984.

[13] C. Decker. Ultra-fast polymerization of epoxy-acrylate resins by pulsed laser irradiation. *Journal of Polymer Science: Polymer Chemistry edition.*, 21:2451–2461, 1983.

[14] J.P. Fouassier and J.F. Rabek, editors. *Radiation Curing in Polymer Science and Technology:Fundamentals and Methods.*, volume 1. Elsevier Applied Science., 1993.

[15] J.P. Fouassier and J.F. Rabek, editors. *Radiation Curing in Polymer Science and Technology:Photoinitiator systems.*, volume 2. Elsevier Applied Science., 1993.

[16] J.P. Fouassier and J.F. Rabek, editors. *Radiation Curing in Polymer Science and Technology:Polymerization Mechanisms.*, volume 3. Elsevier Applied Science., 1993.

[17] G. Odian. *Priciples of Polymerization.* John Wiley & Sons, Inc., 2nd edition, 1981.

[18] L. Flach. Mathematical modelling of a laser-induced photopolymerization process. Progress report, University of Dayton., 1989.

[19] L. Flach and R.P. Chartoff. A process model for nonisothermal photopolymerization with a laser light source. 1:basic model development. *Polymer Engineering and Science*, 35(6):483–492., March 1995.

[20] S. Allanic A.L. Chen, H. Corbel and J.C. Andre. Solid fabrication induced by thermal ir lasers. In *Proceedings of the Third International Conference on Rapid Prototyping.*, pages 3–15. University of Dayton., July 1992.

[21] L. Flach. Personal communications, 1994/95.

[22] L. Flach and R.P. Chartoff. A process model for laser photopolymerization in stereolithography. In *ANTEC 93: Be in that number. Proceedings of the 51st Annual Technical Conference of the Society of Plastic Engineers*, pages 763–766, 1993.

esses created in ceramic shells using Quickcast™ dels

GUE and **P M DICKENS**
rsity of Nottingham, UK

ABSTRACT

Improvements in resins and build styles, coupled with increasing experience, have meant that ever more metal parts are being produced from stereolithography (SL) models via the investment casting route. However, despite these advances, it is still not possible for every foundry to directly use SL models as thermally expendable patterns and gain the same success as achieved with wax patterns.

The central reason behind the inability to investment cast some SL parts' lies in the expansion of the cured resin. The thermoset plastic material of the SL model does not melt during the autoclave process and its expansion creates stresses in the ceramic wall that cause the relatively weak shell to crack.

A work programme is in progress at the University of Nottingham to show *how, why* and *when* these stresses are built up and compare them to the stresses created during the conventional autoclaving of wax parts. The eventual aim of the project is to gain a full understanding of the stresses induced in the models and to develop new build structures that will allow the successful autoclaving and subsequent casting of stereolithography models. Alternative methods of model removal are also investigated.

Details of the work programme are outlined in this paper, along with initial results obtained.

1.0   INTRODUCTION

The use of stereolithography models as thermally expendable patterns in the investment casting process is now an established method of gaining functional metal parts[1]. Many metal duplicates have already been produced from models built in the QuickCast™ 1.0 and the 'new' QuickCast™ 1.1 build styles. However, despite the advances in resin and build styles, the percentage of final castings produced when using SL models falls significantly short of the number achieved when using conventional wax patterns.

Also, the majority of castings that are being produced are being manufactured by specialist investment casting foundries who are willing to experiment and adjust their conventional techniques to accommodate the problems encountered when casting with SL models[2]. Even when successful castings are produced, the dewaxing of the shells (still necessary because of the wax tree that the stereolithography models are mounted on) is generally achieved in the flash furnace. This is despite the fact that the steam autoclave is far and above the most popular method for the

89

dewaxing of shells within todays investment casting industry[3]. There is also evidence to suggest that the autoclave is a more efficient dewaxer than the flash furnace[4]. However, dewaxing shells containing stereolithography models in the autoclave generally results in broken shells[5].

A work programme has begun at the University of Nottingham to investigate the stresses caused by the expansion of the SL models in the green ceramic shells as the construction is passed through the autoclave cycle. From this work, it is hoped that a clearer understanding can be obtained of what is happening to the SL / ceramic system. This should eventually lead to the development of guidelines for the design of 'successful' internal structures.

It is necessary that any new internal structure should allow the shelled models to be autoclaved rather than have the dewaxing realised in the more aggressive flash furnace environment. The reason for this is to open the opportunities for the investment casting of stereolithography models and to allow *any* investment casting foundry to take an SL part and have the ability to achieve a casting. If the shells can survive the autoclave phase, there is generally no problem encountered when the shells are subsequently fired in the flash furnace. Some work has also been performed to investigate alternative ways of dewaxing shells containing existing QuickCast[TM] models

This paper aims to introduce the current work programme giving a general outline and direction research programme. Some initial calculations are also included.

2.0 WORK COMPLETED

2.1 Hollow SL Models

Having autoclaved, without success, a range of models built in the QuickCast[TM]1.0 build style, an experiment was performed that was designed to determine whether it was possible to autoclave even the simplest and weakest of structures. A series of autoclaving trials were carried out on simple shapes with absolutely no internal structure at all, ie hollow parts. This experiment was performed with the rational that if it was *not* possible to autoclave these parts, then it would then be highly unlikely that *any* design of structure would facilitate autoclaving. Conversely, if it were possible to 'dewax' shells containing these hollow SL parts without cracking the ceramic, then there should exist some form of structure in between a hollow and the QuickCast[TM]1.0 structure that will allow the successful autoclaving of the SL pattern.

Hollow spheres, cubes and cylinders were built, treed and shelled (an example of the hollow cylinders is shown in *figure 1*). The shells were then subjected to a standard autoclave at seven bar (165°C) and the dewaxing of all the shells was a complete success. Sectioning of the shells, however, uncovered some unexpected and revealing results.

The 'theory' behind how the QuickCast[TM] models are expected to perform in the autoclave states that as a model expands under the influence of heat (in the autoclave), it should then soften and collapse in on its own voids[6]. If this were the case, on sectioning the autoclaved hollow parts we would have simply noticed a buckling of the SL model away from the walls of the ceramic shell. What *actually* seems to have happened (*figure 2*) is that, far from being intact within the confines of the ceramic shell, the hollowed parts seemed to have gone through some form of thermal shock, and were - in all cases - in several pieces.

What appears to have happened is that instead of the parts expansion inducing failure in the shells, the resins own expansion has promoted the failure of the *part* before it has had time to soften and therefore buckle, as expected. This observation runs contrary to the previously explained reasoning about how the QuickCast™ structure should behave during autoclaving, in that it is designed to soften *and then* collapse in on its own voids. Experience from this experiment alone shows clearly that, in fact, almost the direct opposite has occurred with the models collapsing *before* the resin has had time to fully soften. As the models seemed to have shattered into several pieces within the shells this implies that the parts themselves had 'failed' before they caused failure in the ceramic shell.

2.2 Alternative Methods for Removing Models from Shells

Dewaxing shells containing QuickCast™ models in the autoclave clearly presents problems. The models don't collapse as expected and continue to crack the shells in many cases. Most foundries who are using QuickCast™ models as thermally expendable patterns are achieving the dewaxing of the shells directly in the flash furnace. This can be successful, though foundries are generally adapting their standard techniques to get the parts through this phase as well.

Even though the eventual aim of the research is to develop a build structure to replace the existing QuickCast™ style, work at Nottingham has also been ongoing into ways of removing the current QuickCast™ models by other methods other than the autoclave or flash furnace. The focus of the investigations has been on the use of various liquids and solvents to dissolve and / or disintegrate the resin from the shells. After much experimentation on the effects of various chemicals tested, the choice was originally narrowed down to two chemicals: dichloromethane and nitric acid.

Stereolithography models were treed and shelled in the conventional manner. In order to simplify the problem, the wax tree and runner systems were then largely removed with the use of a hot air gun. This left the stereolithography models encased within ceramic shells ready for treatment with the chemicals. Dichloromethane was poured into one of the shells - with dramatic effect. Instead of the chemical degrading and disintegrating the model, as previously observed, the solvent had an adverse effect on the resin and caused the shell to crack. To further test the reaction of the resin to the dichloromethane, another piece of SL model was subjected to the chemical and filmed with time-elapsed video to observe what happened. The footage clearly shows that the dichloromethane is absorbed into the resin causing the model to expand. The net result of this expansion on shelled QuickCast™ models is that it causes the shells to crack in a similar way to that experienced when placing SL models in the autoclave. It was therefore concluded that the dichloromethane is not suitable for the removal of the resin parts, despite its apparent ability to disintegrate the QuickCast™ models.

Next, the experiment was repeated but with the nitric acid. This time, the shells did not crack and it appeared that the destruction of the model was a success. The shells were then neutralised with baking soda and were fired and cast with aluminium. After casting, the shells were broken off to reveal the casting as shown in *figures 3 & 4*. As can be seen, the resultant casting is less than ideal as it contains many inclusions and imperfections. It is thought that the inclusions have arisen from the initial face coat of the ceramic shell (a fine coat of ceramic that gives definition and surface finish to the subsequent casting) leaking into the unsealed QuickCast™ model. *Figure 4* clearly shows the triangulated QuickCast™ 1.0 pattern within the casting where the face coat leaked into the structure.

Despite disappointment at the poor casting achieved, the removal of the resin model with the nitric acid was generally a success with no deleterious effect on the ceramic shell. The time taken to remove the wax and the model from the shell was significantly higher than the equivalent time to autoclave the part. However, this experiment does show that in the short term, removal of the QuickCast™ model can be successfully achieved with at least nitric acid. The experiment also clearly demonstrates the need to properly seal a QuickCast™ model before shelling.

2.3   Material Properties as a Function of Temperature

As with all materials, the properties of the cured stereolithography resin change with a variation in temperature. The combination of these property changes have a fundamental affect on what happens to the SL / ceramic construction under heating during the autoclave cycle. No accurate data previously existed for the material properties of the epoxy resin at elevated temperatures and a work programme was started to obtain this information.

By determining and contrasting the material properties of the cured SL5170 resin and ceramic materials, it is believed that a correlation between the thermal expansion and mechanical properties can be established that will yield the root cause behind the failure of the shells. The mechanical properties and thermal expansion results are used as the raw data for some stress analysis considered later in this paper, and will be used for the future finite element (FE) simulation of the autoclave phase.

2.4   Thermal Expansion

The expansion of the thermoset SL model within the confines of the ceramic shell construction is the central reason behind the high failure rate encountered during the investment casting process. Clearly its determination is core to the understanding of what is happening under heating during the autoclave cycle. Effectively, the difference in thermal expansion of the ceramic and epoxy materials will yield the strain induced in the system and this is used for future stress calculations.

Figure 5 shows a typical linear expansion graph obtained for the SL5170 epoxy resin. The results were obtained using a linear dilatometer in accordance with *ASTM E228 - 85*. In the range from ambient to about 60°C, there is a linear expansion, α, of approximately 88 x 10^{-6}. In the range from 65 to 150°C there is a twofold increase in thermal expansion to 181 x 10^{-6}. This sharp increase occurs around the glass transition temperature of the cured resin as the secondary bonds of the thermoset plastic material melt.

2.5   Tensile Strength of Resin

Figure 6 gives a resume of the tensile strengths obtained for the SL5170 epoxy resin in the range from 20 to 100°C. The corresponding value for Youngs modulus obtained in the same test is shown in *Figure 7*. The tests were performed in accordance with *ISO 527*. The material was only tested up to 100°C because, as can be seen, the strength (and the corresponding value for Youngs Modulus) is negligible at temperatures at and above 90°C.

The graph clearly shows the sharp decrease in strength of the resin as the samples are ramped through an increase in temperature. This tends to confirm the theory that most of the problems should be occurring at the lower temperatures, where the QuickCast™ model still has enough material strength to not collapse but enough expansion to crack the shells.

2.6   Initial Stress Calculations

To aid in the design of new structures, it is envisaged that eventually the autoclave phase will be simulated using a finite element package. This will be done to predict the failure of the shells for particular structure designs. However, the next stage of the work programme was aimed at producing some results using some basic theory for *solid* stereolithography patterns. This was to help confirm why the SL models are cracking the shells and to predict at what temperature the models will fail for a particular thickness of resin.

For the purposes of this paper, and for simplification of the problem, the calculations are based on the stereolithography / ceramic construction being reduced to an *open ended thick walled cylinder*, with the ceramic 'cylinder' surrounding the resin core (see *figure 8*). When the problem is reduced to this system, it is then possible to apply Lames equations[7] to solve for the stresses in the ceramic shell and the resin core, caused by the systems heating in the autoclave.

The following assumptions are made to further simplify the system

*   The material properties of the ceramic (E_C & υ_C) are considered constant.
*   External pressure is atmospheric.
*   The expansion of the ceramic is considered negligible in contrast to that of the resin.
*   The system is open ended. ($\sigma_z = 0$)
*   Poissons ratio for the materials remains constant.

From Lames equations it can be shown that the pressure at the interface of the surface of the ceramic shell and the stereolithography model, P, can be represented as:

$$P = \frac{\Delta\epsilon}{\frac{1}{E_C}(k + \upsilon_C) + \frac{1}{E_R}(1 - \upsilon_R)}$$

Where:

$\Delta\epsilon$ = Strain induced by expansion of the resin ($= \alpha\Delta T$)
E_C = Youngs modulus of the Ceramic Shell
E_R = Youngs modulus of the resin model.
υ_C = Poissons ratio of the ceramic shell
υ_R = Poissons ratio of the SL5170 resin, and

and

$$k \quad . \quad \frac{(r_C^2 + r_R^2)}{(r_C^2 - r_R^2)}$$

$(r_R \quad = \quad$ Radius of resin 'core')
$(r_C \quad = \quad$ Outside radius of ceramic 'Cylinder')

From the above equation it can be demonstrated that the radial and hoop stresses (σ_r & σ_θ, respectively) in the ceramic at the internal bore of the 'cylinder', (where the maximum stresses occur), are:

$$\sigma_r = kP$$
$$\sigma_\theta = -P$$

and that for the resin core,

$$\sigma_r = \sigma_\theta = -P$$

at any point in the core.

Using the results previously obtained from the mechanical testing of the epoxy resin, values for Youngs modulus and thermal expansion (E_R & α_R, respectively) can be entered in the formula. Poissons ratio for the resin has been approximated from data supplied from Ciba Geigy and is considered to be constant through out the temperature range. Material properties for the shell (E_C & υ_C) have been estimated from data supplied from the National Engineering Laboratory. The values used are:

$E_C \quad = \quad$ 3.5 GPa
$\upsilon_C \quad = \quad$ 0.1
$\upsilon_R \quad = \quad$ 0.4

The thickness of the shell wall ($r_R - r_C$) is constant and is set at an 'average' value of 6mm. Values of hoop stress obtained for resin cores of 0.5mm & 4mm diameter are shown in *figure 9*. Assuming a maximum principle stress criteria, failure of the ceramic shell will occur when $\sigma_{\theta(bore)}$ (hoop stress at the bore of the ceramic 'cylinder') reaches the MOR (modulus of rupture) value of the ceramic shell.

Figure 9 shows that, despite the various assumptions made, the basic profile of the stresses induced seems to confirm what was previously suspected. The stresses in the shell increase and reach a peak as the resin retains some material strength and then rapidly decrease as the material properties of the resin change. The decrease in stress occurs at, or around, the start of the glass transition temperature, T_g of the resin where the material turns from being a true solid to a 'leathery', flexible material with a very low modulus. As the resin core increases in diameter, it is clearly demonstrated that there is rise in hoop stress in the ceramic shell. If the hoop stress equals or exceeds the MOR value of the ceramic, the shell will fail.

Figure 10 shows schematically how the present QuickCast™ structure behaves in the autoclave contrasted with what we are aiming for in any future structure design. Also schematically detailed are a range of shell strengths that broadly represent the present situation.

Any future design of structure must be designed to 'fail' before its expansion causes the failure of the minimum strength shell.

3.0 CONCLUSIONS

Despite reservations about the actual values obtained from the stress analysis equation, due to the uncertainty of the values used for the mechanical properties of the ceramic and the simplified model, it is believed that the trend in stress concentration, seen in *figure 9*, is broadly correct. The peak in hoop stress created by the expansion of the resin at around 60-65°C is consistent with the observations of the author.

Though these results are preliminary and simplified, they clearly show a general trend, and support the idea that the majority of the problems are occurring before or about the beginning of the glass transition temperature of the SL5170 epoxy resin.

Using nitric acid to remove the resin from the ceramic shell was generally a success, and should be considered in the mean time before a 'successful' build structure can be developed to allow the autoclaving of the shelled models.

4.0 ACKNOWLEDGEMENTS

The authors would like to thank 3D Systems Inc. and the Economic & Social Research Council (ESRC) for their help and financial support in this research programme.

5.0 REFERENCES

1   JACOBS, P.F. Quickcast 1.1 & Rapid Tooling. *4th European Conference on Rapid Prototyping & Manufacturing. 13th-15th June, 1995.* pp 1-27. ISBN 0 9519759 4 3

2   DICKENS, P.M. *European Action on Rapid Prototyping (EARP) Newsletter*, No.6, July 1995

3   BREEN, P. The Dewaxing and Firing of Ceramic Shell Moulds. *12th Annual BICTA Conference, Eastbourne, England.* 1973

4   HALSEY, G. Some Factors Influencing the Dewaxing of Ceramic Shell Moulds. *17th Annual BICTA Conference, Bournemouth, England.* May 1970

5   KOCH, M. Rapid Prototyping & Casting. *3rd European Conference on Rapid Prototyping & Manufacturing. 6th-7th June, 1994.* pp 73-76. ISBN 0 9519759 2 7

6   JACOBS, P.F. Stereolithography 1993: Epoxy Resins, Improved Accuracy & Investment Casting. *2nd European Conference on Rapid Prototyping & Manufacturing.* 15th-16th July, 1993. pp 95-113. ISBN 0 9519759 1 9

7   BENHAM, P.P. & WARNOCK, F.V. *Mechanics of Solids and Structures.* pp 331-338 Published by Pitman International, ISBN 0 273 36186 4 .

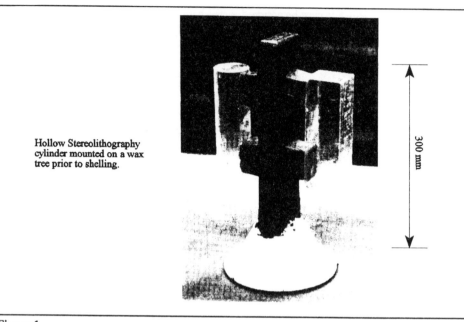

Hollow Stereolithography
cylinder mounted on a wax
tree prior to shelling.

300 mm

Figure 1

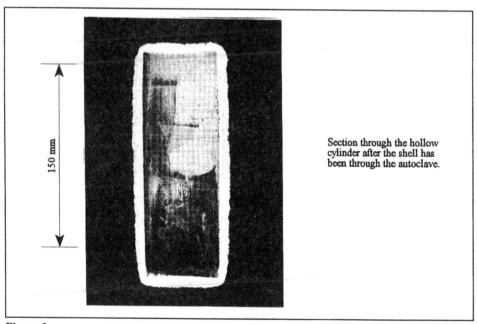

150 mm

Section through the hollow
cylinder after the shell has
been through the autoclave.

Figure 2

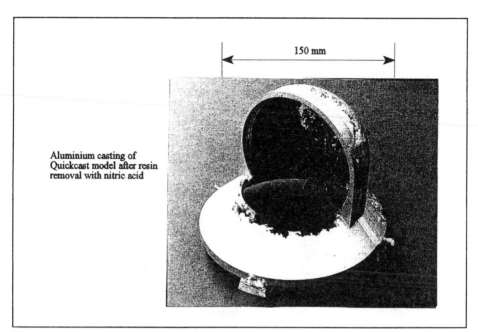

150 mm

Aluminium casting of Quickcast model after resin removal with nitric acid

Figure 3

150 mm

Reverse side of the above casting. Note the inclusions where face coat has leaked into the unsealed model

Figure 4

97

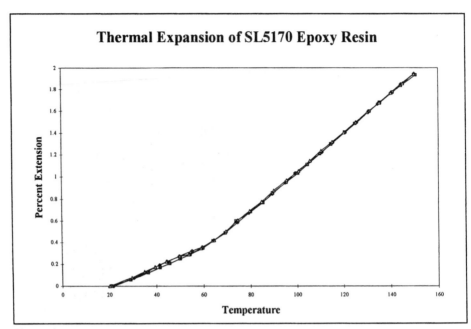

Figure 5

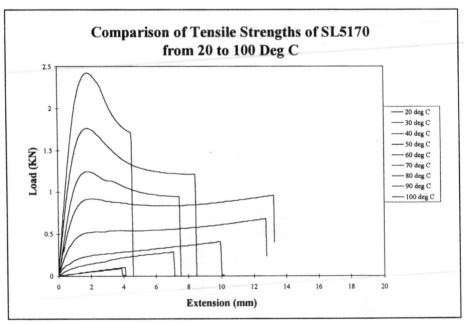

Figure 6

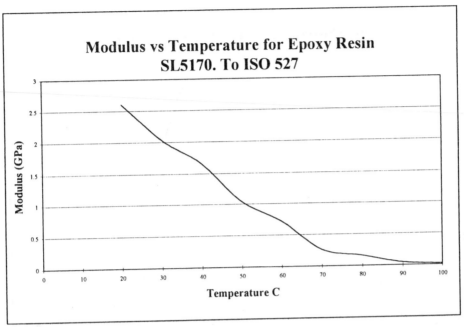

Figure 7

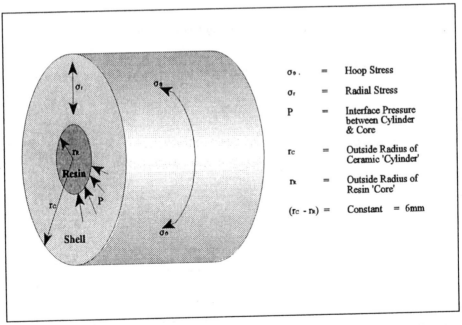

Figure 8

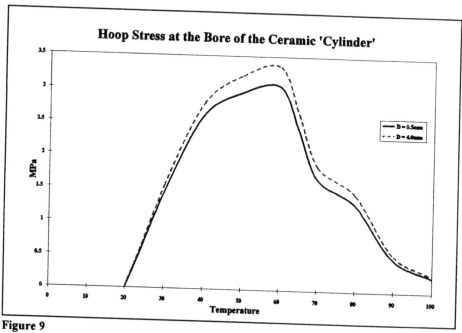

Figure 9

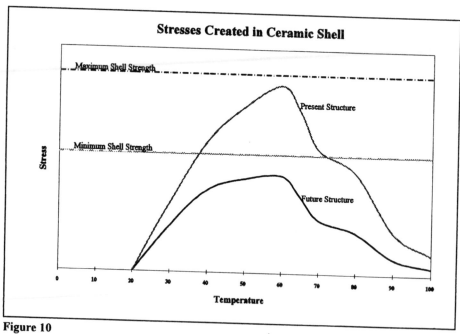

Figure 10

minated object manufacturing: process practice and search experience

REECE
pool John Moores University, UK
J STYGER
vick Manufacturing Group, Warwick, UK

Laminated Object Manufacturing was first commercialised in January 1992, since this time it has become one of the leading and respected rapid prototyping techniques. Indeed many industrial based case studies have been written documenting the benefits from using this rapid prototyping process. However, the process has certain mechanical and physical characteristics which are, as yet, not fully understood. These characteristics include the mechanism of stress creation and the effect of moisture exchange between the models and their environment.

This paper describes the present research work regarding the characteristics of the Laminated Object Manufacturing process being undertaken between the University of Warwick's Warwick Manufacturing Group and Liverpool John Moores University. The paper discusses the various inter-relating factors (such as the thermal, mechanical and material properties) associated with the Laminated Object Manufacturing process and the research being conducted to understand and control their effects, including a mathematical model of the model build process.

1.0 INTRODUCTION

First commercialised in the early 1990's, by Helisys Inc., Laminated Object Manufacturing (LOM) has become one of the leading rapid prototyping techniques. However, it should be noted that there are now other manufacturers, such as Kira Corporation of Japan, either offering or in the process of developing similar systems for commercial exploitation [1]. The work described in this paper has been undertaken on the Helisys equipment using the companies recommended papers.

2.0 DESCRIPTION OF THE LAMINATED OBJECT MANUFACTURING PROCESS

The Laminated Object Manufacturing model is constructed from the base up, on top of a moving platform contained within the working area of the machine. The platform is lowered slightly to allow the material (usually paper coated on one side with polyethylene which acts as a heat sensitive adhesive) to advance into the working position of the system. The platform then rises to allow the material to be bonded to the previous layer of the model. It is then simultaneously compressed and raised in temperature by means of a heated roller which passes over the laminated area (model) and then retracts to a standby position.

A CO^2 laser (25 Watt LOM 1015 and 50 Watt LOM 2030) operated in conjunction with a X-Y plotter arrangement traverses over the bonded laminate to cut the external "box", external and internal profiles of any given cross section that has been generated by the CAD system. The model build sequence is repeated by the platform descent to allow further material to advance into the work area.

Each successive laminate has its nominal thickness calculated each time a layer is bonded in place. This on-line calculation is designed to ensure that, despite the potential of variation in material thickness, the profile which is cut is consistent to the height of the CAD model at that stage of the build process. The calculation is achieved by a feed back system on the Z-axis [2] albeit localised to the area traversed by the feedback roller.

Waste material is created in the working area of the machine during the build process. This material is not required as a part of the finished model and is therefore sliced into cubes by the laser. The cube formation is a deliberate part of the procedure and allows the removal of the waste from the model perimeter once the build process is completed. The cubes are intended to provide support to the model during the compression operation. This technique ensures that the roller passes over a complete sheet of laminate and that the model sections are adequately supported during the cutting operations. Figure 1 illustrates the layout of the Laminated Object Manufacturing systems.

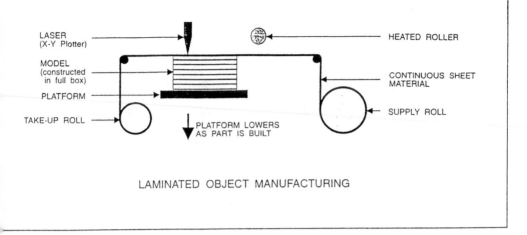

LASER (X-Y Plotter)
HEATED ROLLER
MODEL (constructed in full box)
CONTINUOUS SHEET MATERIAL
PLATFORM
TAKE-UP ROLL
SUPPLY ROLL
PLATFORM LOWERS AS PART IS BUILT

LAMINATED OBJECT MANUFACTURING

Fig. 1

Once the build process is completed the model is removed from the machine as a complete block. The perimeter box and support cubes are removed to expose the finished model.

3.0 THE LEVELS OF ACCURACY

Consistent part accuracy must be one of the cornerstones of successful rapid prototyping. Due to the layer build technique, it is possible to consider two possible areas of error regarding the build process, these are;

1. Linear errors which consist of :
   a.   errors in x-y dimensions
   b.   errors in z height
2. Warpage and curl

Warpage and curl will not be covered in this paper.

3.1 Possible Errors In x-y Dimensions

In the Laminated Object Manufacturing process, the laser follows the true line of data (i.e the line of the CAD model). The machines software does not house a beam compensation routine and this can potentially lead to errors in the model build procedure. The quoted laser diameter is 0.010 ins - 0.015 ins (0.254 - 0.381mm) [3]. It is therefore possible for a Laminated Object Manufacturing model to be either undersized or oversized by 0.005 ins - 0.007 ins (0.127 - 0.177mm) on each surface depending on the geometry. Figure 2 illustrates the potential areas of error assuming there is no beam compensation routine. Several recommendations have been made regarding the use of globally scaled data to compensate for the beam. However, this will only succeed on certain parts of the build, because internal and external features will require opposite scaled compensation factors.

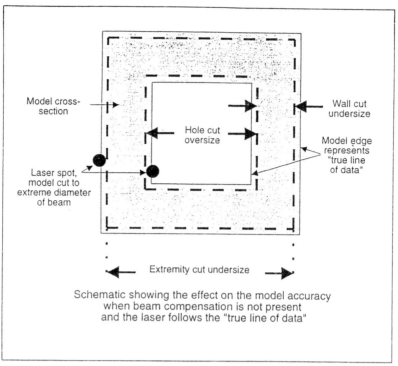

Model cross-section

Wall cut undersize

Hole cut oversize

Model edge represents "true line of data"

Laser spot, model cut to extreme diameter of beam

Extremity cut undersize

Schematic showing the effect on the model accuracy
when beam compensation is not present
and the laser follows the "true line of data"

Fig. 2

3.2 Errors In z Height

It has always been recognised that Laminated Object Manufacturing models expand in the z direction [4]. Indeed Helisys have made recommendations that the models should be built 2% undersized to counteract this phenomena [5]. The expansion has traditionally been associated with moisture absorbtion at a point after completion of the model build cycle. However, initial tests have indicated that the models are typically expanding much sooner than anticipated. It has also been noted that the models are expanding in the z direction at significantly different rates, typically between 1.56% and 8% [6].

A series of experiments have been carried out to determine the causes of z height expansion in the models. The experiments have been based around simple cubes of known dimensions. This eliminates any potential error in STL file translation and topography blending. By monitoring the temperature/time factors immediately after completion of the model build cycle, it has been possible to establish a curve relating to z height movement. This curve cannot be solely reasoned in favour of simple moisture exchange.

The examples below are all based on a series of 50mm test cubes, which were measured on a Cordax 1808 DCC coordinate measuring machine fitted with a Renishaw PH9 Mk2 head. Figure 3 illustrates the comparison of temperature and time (12 hours) against model height. It can be noted that the models are built in a heated environment in the region of 60°C and that the models typically show a tendency to contract until they reach a state of equilibrium with ambient room temperature (approximately 30°C). The thermal nature of the process is not taken into account regarding any suitable compensation software. It is therefore reasonable to assume that the models will become undersize as they contract after the build has been completed. However, it is interesting to note that the models are not typically built at the correct z dimension. A similar phenomena has also been recorded for measurements in the x-y plane.

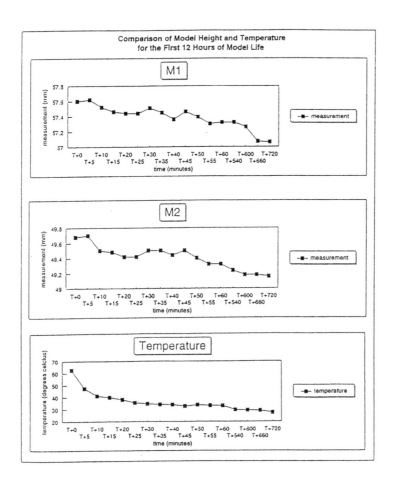

M1 = measurement from base of model build table (ie model + tiles)
M2 = measurement of model from commensurate of build (ie model but no tiles)

Fig.3

To date, the experiments have shown that the parts expand to nominal size at T+30 hours. The experiments have also indicate that identical models built as a group, at the same time and stored in the same environment will show a tendency to expand at different rates. This phenomena is illustrated in Figure 4, which shows the long term expansion characterised of a group of six 50mm test cubes.

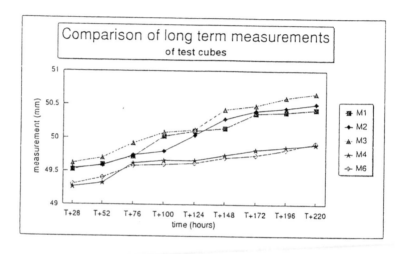

Fig.4

The experiments have also indicated that there will be a correlation between the level of expansion and the position of the model on the build table in relation to the other neighbouring parts. Larger test cubes have also demonstrated deviation across the planar surface. It has not been possible to simulate the rate or level of expansion of the model build material out side the environment of the machine. This has lead to the conclusion that although moisture does effect the expansion of the Laminated Object Manufacturing part (typically in the region of 0.1%), it is typically a small contributing factor compared to other mechanical influencers [7].

4.0 THE INFLUENCING FACTORS ON MODEL BUILD

The experimental work to date has indicated that the Laminated Object Manufacturing models are influenced by several factors which affected their immediate and long term stability. Figure 5 illustrates the five main influencing factors affecting the model build accuracy.

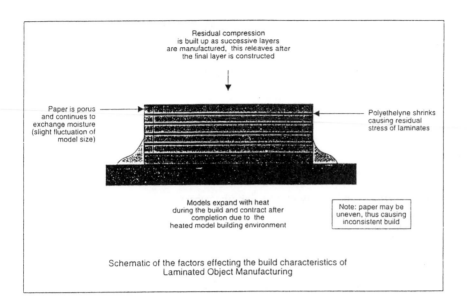

Residual compression
is built up as successive layers
are manufactured, this releaves after
the final layer is constructed

Paper is porus
and continues to
exchange moisture
(slight fluctuation of
model size)

Polyethelyne shrinks
causing residual
stress of laminates

Models expand with heat
during the build and contract after
completion due to the
heated model building environment

Note: paper may be
uneven, thus causing
inconsistent build

Schematic of the factors effecting the build characteristics of
Laminated Object Manufacturing

Fig. 5

Each factor typically co-exists and inter-relates to the others during the model build process. However, from the work undertaken to date, it is possible to place the groups according to major effect on accuracy and stability as listed below;

1.   Compressive load caused by the heated roller during bonding
2.   Interlaminar stress and bonding
3.   Beam compensation
4.   Moisture exchange
5.   Inconsistent model build material*

(* Factor 5 is variable and could influence any of the above to a greater or lesser degree).

5.0 PREDICTIVE ANALYSIS

The early experimental work has demonstrated that there are a number of complex, interrelated factors which affect the model build process and have not yet been quantified to the complete satisfaction of the authors. These difficult to control parameters contribute to the unpredictable nature of the process. However, it is generally considered that some of the challenges of the process could be overcome by understanding the build procedure and developing a predicative model of the machine's characteristics.

Initial work has provided evidence of the temperatures attained within the process and it is believed that the pressure of the roller and the temperatures achieved in bonding are the two main factors associated with deformation and stress creation. The work currently being undertaken is designed to produce a mathematical model of the Laminated Object Manufacturing process in order to achieve consistently accurate parts from every build.

Initial efforts at modelling the temperatures involved have been attempted using the equations given by Carslaw and Jaeger [8] where:

$$t = \frac{q'}{2 \pi K} \, e^{(Ux/2\alpha)} \, K_o \, [U \, (x^2 + z^2)^{1/2} / 2 \, \alpha]$$

where :

> q' = Strength of the heat source per unit time per unit length along the y-axis.
>
> K = Thermal conductivity of heat sink (the paper).
>
> α = Diffusivity of the heat sink material
> $= \dfrac{K}{\rho \, c}$
>
> c = Specific Heat Capacity of the heat sink material.
>
> U = Velocity of the source (the roller) parallel to the x-axis.
>
> $K_o(x)$ = Modified Bessel Function of the second kind of order zero.
>
> x,z = Locations within the heat sink material.

5.1 An Experimental Case Study

From the text we have the equation for the temperature at any point in a medium subjected to heating from a moving line source. This is given as :

$$t = \frac{q'}{2 \pi K} \, e^{(Ux/2\alpha)} \, K_o \, [U \, (x^2 + z^2)^{1/2} / 2 \, \alpha]$$

(see above for values)

It is clear that the temperature achieved is dependant upon not only the properties of the materials involved but also the velocity of the source of heat.

We expect a temperature difference of around 90°C over 5mm due to the insulating nature of the material. Inserting suitable figures into the equation :

Considering a point on the centre-line of the source and 5mm below the contact area

$$90 = \frac{q'}{2 \pi \, 0.5954} \, e^{(0.01 \, x \, 0/2 \, x \, 7.3e\text{-}8)} \, x \, K_o[0.01 \, x \, (0.005^2 + 0)^{1/2} / (2 \, x \, 7.3e\text{-}8)]$$

$$= \frac{q'}{3.741} \, x \, 1 \, x \, K_o \, [342.46]$$

This yields a value for the Bessel function of zero.

Considering instead a point 5mm in front of the source and on the surface of the paper :

$$90 = \frac{q'}{2 \pi \, 0.1784} \, e^{(0.01 \times 0.005/2 \times 2.3e-7)} \times K_o[0.01 \times (0+0.005^2)^{1/2} /(2 \times 2.3e-7)]$$

$$= \frac{q'}{1.1209} \, e^{4.954e-5} \, K_o \, [108.7]$$

This also yields a value for the Bessel function of zero.

The solution from Carslaw and Jaeger is a true analytical solution to the problem of heat transfer from a moving line source under steady conditions. The steady case can be used since although the heat sink material is moving in relation to the source, the heat affected zone will maintain the same shape, size and temperature during the entire heating period.

Prediction of the temperature at a point in the heat affected zone is difficult in the case of the LOM model because :

a) The paper material is an excellent insulator for the conduction of heat;

b) The layered structure of the model makes it difficult to determine exact values for conductivity and diffusivity and, as such, mean values must be used.

The combination of these factors causes the equations of Carslaw and Jaeger to produce incorrect results. This is an important point since it echoes the problems experienced in understanding the building process in Laminated Object Manufacturing and the challenges to be overcome in modelling the process.

It was expected that the calculation would provide a temperature difference of zero in contact with the source, becoming greater as the distance from the roller increased. A temperature of around 90°C would be expected over 5mm due to the insulating nature of the material. However, this has not proved to be the case in practice since the Modified Bessel Function tends to zero at all but the smallest distances from the source (i.e. the heated roller).

5.2 An Alternative Method of Predictive Modelling

The alternative method of modelling that has been proposed is that of using a computational fluid mechanics package (FIDAP) to model the process. The model will take account of the pressure of the heated roller and of the heat being transferred from the roller into the paper material. It should be recognised that this work is still in its early stages of development, however, simple correlations have been achieved on test samples.

6.0 CONCLUSION

Laminated Object Manufacturing has become established as a "proven" rapid prototyping technique with an industrial user base. However, the process is as yet not fully understood by the users or monitored within the machine environment. The apparent lack of academic references on the subject suggests that a better scientific understanding of the system is imperative if the technology is to continue to develop. The work conducted to date has enabled a much clearer understanding of the build process to a point where it is now possible to predict the behaviour of certain geometries under certain conditions. It is envisaged that this work will ultimately lead to a full predicative tool for Laminated Object Manufacturing users.

REFERENCES

[1]   "Research into Rapid Prototyping". David Wimpenny, Rapid News, Volume 3 Number 1, 1995. Page 3-7.

[2]   "Helisys Laminated Object Manufacturing". L Bowman, Proceedings from 1st European Conference on Rapid Prototyping, University of Nottingham, 6-7 July 1992. Page 59-63.

[3]   "The Power of LOM is Now Within Reach". Anon., Technical literature, Helisys Inc., 2750 Oregon Court Building M-10, Torrance, CA 90503, USA.

[4]   "Helisys Users Group - Annual Meeting". Paul Hutchinson, A report submitted by Webster Mouldings Ltd on the Helisys User Group Convention , Chicago, USA, 31 May - 2 June 1995. Page 4.

[5]   "Laminated Object Manufacturing Manual". Anon, Published by, Helisys Inc., 2750 Oregon Court Building M-10, Torrance, CA 90503, USA.

[6]   "The Influences Effecting The Stability And Accuracy Of Laminated Object Manufacturing Models". L E J Styger, a submitted thesis by the author to the University of Warwick, 31 August 1994.

[7]   "The Study of Stress and Deformation Within the Laminated Object Manufacturing Process". S R Reece et al, submitted for the AMPT '95 Conference, Dublin University, August 1995.

[8]   "Conduction of Heat in Solids". H S Carlslaw, J C Jaeger, Oxford University Press, Second Edition, 1980. ISBN 0 1985 3303 9.

L E J Styger   S R Reece
Advanced Technology Centre   Liverpool John Moores University
Warwick Manufacturing Group   SETM
University of Warwick   Byrom Street
Coventry   Liverpool
CV4 7AL   L3 3AF

30 June 1995

reolithography process improvement

HMATI and **P M DICKENS**
rsity of Nottingham, UK

1.0 SYNOPSIS

Stereolithography(SL) has attracted the most attention of all the Rapid Prototyping techniques. However, the stereolithography method is still restricted in some of its applications due to the lack of accuracy and theoretical process knowledge available. Previous studies indicate that there is great potential for further improving accuracy and reducing part distortion.

Currently, much research is concentrated on different aspects of accuracy, which is mainly a function of shrinkage and distortion during polymerization. A research project has started with the aim of understanding the SL photo-polymerization process in order to improve its accuracy and efficiency.

Experimental tests have been developed aimed at better understanding the SL process, i.e. single, double, and triple layer samples are built with varying hatch spacings. Three different approaches were used to study and analyse the samples, namely: Scanning Electron Microscope (SEM), Profilometry, and Sectioning.

2.0 INTRODUCTION

A problem with stereolithography is part inaccuracy and distortion which is the result of resin shrinkage during photo-polymerization. Shrinkage is an inherent property in polymerisation where the average distance between monomer groups decreases and results in an increase in density. Shrinkage can cause internal stresses in the model and different parameters are attributed to shrinkage, such as laser exposure energy(1), polymerization rate (2), build style (3), degree of polymerization, the polymerization kinetics, etc. Ultimate shrinkage and shrinkage rate are dependant upon the amount of energy used to polymerize the polymers (1). Faster shrinking resins exhibit lower overall shrinkage than slower shrinking resins, because less shrinkage occurs after the strand has been scanned (2).

During build time, material properties and shrinkage change continuously and its rate of change decreases with time(2). However this change can be divided into two phase, instantaneous and long term shrinkage. Instantaneous shrinkage is the changes, the polymer undergoes during polymerisation and long term shrinkage is the ongoing changes which takes place after the model is built.

Advanced build styles such as STARWEAVE™ have reduced warpage and dimensional inaccuracy significantly (3). STARWEAVE™ was developed by 3D Systems during 1990 and 1991. WEAVE was intended to minimise the fraction of residual liquid resin within the part. It builds each layer in incremental stages. It first draws the X-hatch vectors and as a result partially polymerizes the layer. This allows the layer to shrink independently from the rest of the model, it is then attached to the model by drawing the Y-hatch vectors. STAR is a further improvement of this build style, it is an acronym for Staggered hatch, Alternate sequencing, and Retracted hatch from the borders. While WEAVE was very successful in reducing postcure distortion, STAR was aimed at reducing internal stresses and distortion and further improving the accuracy. However shrinkage is an ongoing process and STARWEAVE™ has only been able to reduce the initial polymerization shrinkage. The influence of alternate staggering of layers on curl is not significant. In general a higher magnitude of curl distortion is observed in the weave build style (5) (6).

Layer thickness directly controls the build cycle time (5). Layer thicknesses of below 0.12mm, consume a majority of time in the recoating process. While for layer thicknesses above 0.25mm, the build cycle efficiency drops because the laser scan velocity decreases in order to maintain the desired layer thickness. In general as layer thickness decreases, optimum hatch spacing gets smaller and the dependency is the square root of layer thickness. Figure 1 shows a schematic diagram of some build parameters. As hatch spacing goes above the optimum value, it results in a softer material. As hatch spacing goes below the optimum value, it results in an overcured part and introduces more curl and distortion. Hatch overcure values must be based on a balance between curl distortion and layer delamination. Skin fills are used to seal the part. They are equivalent to putting the final sheet metal on an airplane frame. Previous studies indicate that the values of fill cure depth between 0.0mm and 0.63mm produce good results (7). Therefore for the type of resin, and laser spot size, the optimum hatch spacing must be determined for different layer thicknesses.

The purpose of the present experiment was to look at the actual behaviour of the STARWEAVE™ build style technique while applying it to a development acrylate based resin from Zeneca. Acrylate resins polymerise by free radical mechanisms, causing a very rapid process with a relatively high shrinkage of 5-6%. Upon laser irradiation the polymer chains are formed as a result of reactive double bonds which are polymerised (4). The rate of polymerization depends on the type of photoinitiator used and on its concentration, and functionality of both the photoinitiator and the monomer.

3.0 EXPERIMENTAL

An SLA-250 stereolithography machine was used to build test parts with Zeneca 93/9 acrylate resin. Test parts were simple boxes of 20mm x 20mm. Layer thickness was 0.25mm. Due to the nature of the STARWEAVE™ build style, hatch spacing values were chosen to generate samples with sufficient green strength. Hatch spacing was varied from 0.20mm to 0.26mm with a constant layer thickness of 0.25mm. All test parts were built in single, double, and triple layers. They were each identified e.g. 0.20-1, where 0.20 is the hatch spacing in millimetres and 1 is the number of layers. All test parts were built in two sets, the first was used for sectioning while the second was used for profilometry and SEM experiments.

3.1   Methods:

In order to study and analyse the photo-polymerization process experimentally, different approaches may be applicable. Three methods which were identified as being suitable are as follows :

1- Sectioning
2- Scanning Electron Microscope (SEM)
3- Profilometry

These three methods were applied in order to study :

1.	Scanning behaviour	7.	Retractions
2.	Alternate Staggered hatching	8.	Hatch spacings
3.	Gaussian laser profile	9.	Line width
4.	Max and min layer thickness	10.	Layer thickness
5.	Maxima at X and Y hatch intersections	11.	Cavities
6.	Hatch overlap	12.	Trapped resin

3.1.1   Method 1: Sectioning

The sample polymer was carefully prepared to reveal the section without it being damaged, or deformed. Sectioning, mounting, grinding and polishing were necessary to prepare the sample for examination. This revealed many things including hatch spacing (hs), line width (Lw), local maxima at X and Y alternative hatch intersections, alternate retraction, staggered hatches, layer thickness, strand cross section size/shape, uncured trapped resin (residual), cavities, and internal defects. From the samples it is possible to see the retraction distance very clearly. To avoid any damage to the samples, room temperature curing resins were used. EPO-KWIK resin cures in 30-40 minutes with maximum temperature of 40°C (8). For each specimen, in excess of 100 measurements were taken, an example of which is shown in Figure 2.

This was a difficult and time consuming process. Variations from the STARWEAVE™ build style indicated problems with the stereolithography process. But, in spite of all these problems, in general this method has inherently a high potential of answering most of the queries in studying the behaviour of photo-polymerization process. It is the only method which looks at the interior cross section of the samples where most of the answers lie.

3.1.2   Method 2: Scanning Electron Microscope (SEM)

Before placing the specimen in the SEM chamber, the sample was carefully gold coated. Gold Sputter Coating gave 2-3 angstroms of gold, in order to make the sample conductive. SEM enables observation and accurate measurement of external features.

3.1.3   Method 3: Profilometry

Profilometry appears to be a good method to look at the hatch profile. It is very quick, and relatively accurate. The equipment used for this purpose was a Surfcom 30C. It is an accurate and versatile surface measuring instrument which indicates both centre line average roughness (Ra) and root mean square roughness (Rs) and records roughness, waviness and total profiles. However, in this experiment our prime intention was to look at the total profile of the sample surface from underneath. This profile revealed the relative hatch spacing, the local maxima and minima, and in general identified the hatch behaviour.

The surface profile was sensed with a differential transformer fitted with a diamond-tipped stylus. A stylus tip radius of 3μm allowed very fine surface irregularities to be gauged accurately while a stylus force of 0.5 gr enabled the delicate surface of the polymer to be gauged without damage.

4.0 EXPERIMENTAL RESULTS

Both set of samples were investigated by applying sectioning, SEM, and profilometry methods. The results from these three methods are in accordance with each other. Based on the machine settings, and considering the STARWEAVE™ build style, the following observation were expected to be seen :
▸ Clear image of hatches
▸ Alternate staggered hatches
▸ Alternate retraction from boarders
▸ Gaussian profile of cured polymer
▸ Linewidth size
▸ Hatch overlapping
▸ Local maxima at X and Y orthogonal intersections
▸ Layer thickness
▸ Trapped resin and cavities

4.1   Hatch spacing of 0.26mm:

Layer thickness   varied from 0.216mm to 0.26mm, while it was expected to vary between 0.225mm and 0.3137mm. These two values correspond to the X hatch layer thickness and Y hatch layer thickness. In other words the second value is supposed to be the depth of the local maxima at vector intersections. Retraction distance varied between 0.05mm and 0.1mm, and was also irregular. Surprisingly in some cross sections retraction was observed from both borders, and sometimes retraction did not alternate at all. According to STARWEAVE™, the retraction must be repeated alternatively in both axis. There may be two reasons why this did not occur. First, the Linewidth compensation was wrong. Second, the retraction value must be modified for the Zeneca resin.

Hatch spacing was 0.254mm instead of 0.26mm (Figures 3, and 4). Layers were partially staggered by about 41.5%, or 0.076mm instead of 0.13mm (See Figure 4). Hatch spacing was between 0.26mm and 0.267mm for sample 0.26-3. There were no clear peaks (maxima) at X and Y hatch intersections (See Figure 5, 6, 7, and 8). No cavities were observed between strands or layers for sample 0.26-3. However, cavities were clearly seen in samples 0.26-2 (See Figure 6) and 0.26-1 . The top surface was nearly flat for sample 0.26-3 (See Figure 9, and 10).

4.2   Hatch spacing of 0.24mm:

Similar observations were seen with 0.24mm hatch spacing, as for 0.26mm hatch spacing. Layer thickness varied from 0.229mm to 0.26mm, where the minimum thickness was very near to the estimated value of 0.225mm while maximum layer thickness was expected to be 0.3137mm. As before no local maximas were observed. Retraction distance varied between 0.025mm to 0.095mm, and was also irregular. However, there were some hatches without retraction. Layers were partially staggered by 41.5%, or 0.07mm, instead of 0.12mm, which is the same as 0.26mm samples. No gaps or cavities were seen between strands or layers. The top surface was nearly flat, and there were no gaps between the strands on the top surface which means the linewidth is near to the hatch spacing.

4.3   Hatch spacing of 0.22mm:

Layer thickness varied from 0.229mm to 0.26mm. The top and bottom surface was almost flat (Figure 11). The hatches were so close that a clear image of hatch profile cross section was not visible. There was not always complete retraction (Figure 11, and 12). Hatch spacing was 0.21mm and not 0.22mm. There were no local maxima observed.

4.4   Hatch spacing of 0.21mm:

Layer thickness varied from 0.229mm to 0.26mm. The top and bottom surfaces were flat (Figures 14, and 15). There was no retraction whatsoever. As hatch spacing was reduced, it became more difficult to measure features such as local maxima, linewidth, etc. This was because of the large overlap with individual scans.

4.5   Hatch spacing of 0.20mm:

Layer thickness varied from 0.267mm to 0.3mm which was above the estimated values of 0.225mm and 0.3137mm respectively. The bottom surfaces were flat and no retraction was seen.

5.0 DISCUSSION

5.1   Hatch Spacing & STARWEAVE™

As hatch spacing decreased, minimum cure depth ($C_d(1)$) was increased, while maximum cure depth ($C_d(2)$) was nearly unchanged. This may be due to the fact that as hatch spacing gets smaller or in other words strands get closer to each other, the portion of overcured hatch gets larger and therefore it results in an increase in cure depth between the strands. However, because parameters affecting $C_d(2)$ are not changed (i.e layer thickness a, critical exposure Ec, depth of penetration Dp, laser power Pl, etc), the maximum cure depth which is the result of X and Y orthogonal intersections remain nearly unchanged.

Characteristics of STARWEAVE™ are well presented in ref. (3), however in practice we have seen variations from this build style. In fact, none of the samples showed all of the standard features of the STARWEAVE™ draw style. Local maximas at vector intersections are responsible for the nth layer to attach to the $(n-1)$th layer. These local maximas are claimed to act like tiny rivets and prevent delamination if properly set. According to STARWEAVE™ the minimum cure depth expected is :

$$\text{Min Cure Depth} = C_d(1) = \text{Layer Thickness 'a'} - 0.0254 \tag{1}$$
$$\text{Max Cure Depth} = C_d(2) = a + [0.6931\ Dp - 0.0254] = C_d(1) + 0.69\ Dp \tag{2}$$

Where $C_d(1)$ is the cure depth in the first pass, and $C_d(2)$ is the maximum cure depth at orthogonal intersections and Dp is the resin depth of penetration. Jacobs (3), has shown that where lasers have non-Gaussian irradiance distributions there is very little effect on the working curve, and hence on the value of Cd, but there will be a somewhat greater influence on the value of Linewidth (Lw). This is more important on the SLA-250 where a helium-cadmium (He-Cd) laser is used. This laser is often multimode and may even occasionally have a "doughnut-like" irradiance distribution, when the laser is nearing the end of its life. The laser on the SLA machine at the time of these experiments was in fact at the end of its life. This may have caused a change in the shrinkage if the amount of energy is also affected, and a change in the size of the actual cured Linewidth which is predicted by following equation.

$$Lw = B\sqrt{(Cd/2Dp)} \tag{3}$$

where B is the laser spot diameter, Dp is the penetration depth, Cd is cure depth, and Lw is the cured linewidth. However further experiments after the new laser installation showed the same result with respect to the laser beam profile (Figure 13). The purpose of alternate staggered hatch is to avoid weak spots (3), and reduce stress concentrations along the relatively weaker regions between vectors. Therefore the nth layers are offset by exactly half the regular hatch spacing, relative to those on the $(n-1)$th layer. However, as seen from samples, they do not always exhibit the alternate staggered hatch pattern completely (Figure 4). Instead the hatch pattern is alternatively staggered by about 59%. This may affect the STARWEAVE™ process, and intensify the stress concentration of the model, leading to additional curl distortion.

It was noticed that some of the samples were deformed at corners, which may be attributed to the lack of retraction and/or continuous retraction of strands from one side. When retraction was introduced to STARWEAVE™, it was aimed to introduce free shrinkage and lessen the problem of stress concentration and to improve the flatness of slabs (3). This non-uniformity in retraction can affect the overall part accuracy. The top surface profile of samples 0.26-3 and 0.21-2 are shown in figure 9, figure 10, figure 14, and figure 15, respectively where there is a distinct difference in the X and Y directions.

5.2   Induced stress

It was observed that hatches in one axes were properly built but in the other axes were irregular. It first appeared that something was wrong with the process. However later observation revealed the fact that when the first pass is drawn, it's thickness $C_d(1)$ is 0.0254mm less than the layer thickness. This means the first pass does not touch the layer below and will have a free shrinkage. As the second pass is drawn with the same exposure, each strand undergoes a change in the cured cross section at the intersection of strands (see figure 13). This difference in cured cross section gives a nonuniform shrinkage and therefore produces stresses in the structure which is then under tension. This takes time to reach a relieved state.

In addition, previous studies indicate that a percentage of the acrylate monomers remain unreacted within the part and trapped radicals continue to react slowly over time and therefore intensify the internal stress (12). Figure 13 clearly shows the difference between the first and second pass. Of course one can say that first and second pass are alternately changing, but the fact is that by modifying the present build style such as reducing the exposure in the second pass, it may reduce the biased effect of stress towards one direction and result in a more uniform distortion, and less overall stress in the structure.

5.3   Laser Beam issues

The precisely defined exposure conditions that arise from the use of laser sources, demand a deeper understanding of the laser beam interaction and polymerisation process in particular. Lasers in practice tend to emit a beam which is a superposition of 'modes' (11). The narrowest, least divergent, of these modes has the properties of a so called Gaussian beam, which usually refer to the properties of an ideal beam (Figure 16-a). The present SLA 250 beam profile output shows that the form of the beam is not quite Gaussian and may be changing to a multimode state(Figures 17-a, 17-b, and 13). When the laser power is fixed, the beam cross section profile indicates its uniformity. Therefore any deviation from a parabolic profile results in a loss of maximum cure depth, which all calculations are based on. However, the laser beam quality may be increased by adjusting the aperture at the beam source (9). This will increase the laser efficiency in order to get a more homogeneous intensity distribution.

Figure 16 shows a sketch of the intensity of two profiles, case 'a' is the ideal Gaussian beam "TEM_{00}" and case 'b' is a laser beam with a large amount of the laser energy emitted in the next-higher mode "TEM_{01}" (11). Case 'b' is about 50% wider than the TEM_{00} mode, and as a result 50% more divergent. Figure 17 indicates two types of intensity profiles in practice, consisting of a mixture of the cases in Figures 16-a and 16-b. The beam diameter for the purpose of calculation is taken at the $1/e^2$ or 13.5% of the peak intensity Ho. Although the beam energy outside this diameter is far less than Ec and practically should have no effect on polymerisation, some photos reveal that   polymerisation reaction   may be initiated. This quasi-photo-polymerisation may facilitate the reaction when the resin is exposed for the second time. Figure 18 shows the cross section in the X axis of the double layer sample with its support. Therefore the Y strand is drawn from left to right and as it gets to the support area where it has just been exposed during the first pass, cure depth goes beyond 1.5 to 2 times the layer thickness. This may be attributed to the quasi-polymerisation reaction just near to the supports . One may say that it is due to the residual resin which is cured during postcuring, but it is observed that this phenomenon at the support walls is alternating due to the retractions.

Based on the above argument the change in hatch cross section profile from Gaussian to doughnut shape may be attributed to the age of the laser. However, further experiments with the new laser gave the same result with respect to the laser beam profile. Other problems such as nonuniform retraction, staggered hatching may be attributed to the control algorithm, galvanometer, or incorrect settings of the apparatus.

6.0 Conclusions

- Results show that the build parameters such as retraction, cure depth, staggered hatching and etc. are not consistent. This could be due to the variation in machine, process, or material parameters such as laser power, recoating mechanism, resin parameters etc which may impose a change in linewidth, layer thickness, staggered hatch,etc.

- The first pass is dominated by the second pass. When the second pass is drawn, the first pass is hidden and as a result, the stress created in the structure as a result of restricted shrinkage, may be directed towards the second pass where the changes are taking place.

- The free shrinkage during the first pass takes place in all three dimensions which results in displacement in the X, Y, and Z axis. The shrinkage during the 2nd pass is responsible for the induced residual stress. Shrinkage at the intersections is different due to the restriction of shrinkage between the intersections which have been previously polymerised during the first pass.

- After each single strand is finished the laser is still working and directing photons at the resin surface. The scaning speed increases from the end of one strand to the start of the next. This should mean that the laser energy is not enough to cure the resin. However, some curing does occur. Therefore the new strand is the continuation of the last strand, and a full retraction is not quite achieved from the borders (see Figures 11 and 12).

- The cured polymer profile indicates (See Figure 13) a problem with the laser which is not exhibiting exactly the Gaussian laser beam profile. This may distort the intensity profile and result in a different cured cross section.

- In STARWEAVE™, the connection between layers is expected to be through local maximas. While in practice layers are connected through almost flat surfaces at the bottom of each layer which is equivalent to the over cure thickness of 0.0637mm in STARWEAVE™. This may affect the overall distortion.

- The percentage of the unreacted acrylate monomers within the structure which will continue to react slowly over time may be responsible for the 'long term' shrinkage and induced stress.

Acknowledgements

The authors would like to thank ZENECA for their help and financial support in this research programme.

REFERENCES

1.   Weissman, P. T., Bolan, B. A. and Chartoff, R. P."Measurements of linear shrinkage and the residual stresses developed during laser photo-polymerization ",The 3rd International

Conference on Rapid Prototyping, Edited by Marcus, H. L., University of Dayton, Dayton, June 7-10, 1992, pp 103-112, ISSN 1053-2153.

2. Flach, L., Chartoff, R. P."A simple polymer shrinkage model applied to stereolithography", Solid Freeform Fabrication Symposium, Edited by Marcus, H. L., University of Texas at Austin, Austin, Texas, Aug 8-10, 1994, pp 225-233, ISSN 1053-2153.

3. Jacobs, Paul F."Rapid Prototyping & Manufacturing, Fundamentals of Stereolithography", Dearborn, MI: Society of Manufacturing Engineering, 1992, ISBN 0-87263-425-6.

4. Decker, Christian"Kinetic analysis of laser-induced reactions in polymer films" SPIE Laser-Assisted processing II, 1990, Vol. 1279, pp 50-59.

5. Jayanthi, S., Keefe, M. and Gargiulo, E. P."Studies in stereolithography: influence of process parameters on curl distortion in photopolymer models", Solid Freeform Fabrication Symposium, Edited by Marcus, H. L., University of Texas at Austin, Austin, Texas, Aug 8-10, 1994, pp 250-258, ISSN 1053-2153.

6. Ullett,J. S., Chartoff,R. P., Lightman, A. J., Murphy,J. P. and Li, J."Reducing warpage in stereolithography through novel draw styles", Solid Freeform Fabrication Symposium, Edited by Marcus, H. L., University of Texas at Austin, Austin, Texas, Aug 8-10, 1994, pp 242-249, ISSN 1053-2153.

7. Horton, L., Gargiulo, E. P. and Keefe, M."An Experimental Study of the Parameters Affecting Curl in Parts Created Using Stereolithography", Solid Freeform Fabrication Symposium, Edited by Marcus, H. L., et al., University of Texas at Austin, Austin, Texas, 1993, pp 178-184, ISSN 1053-2153.

8. BUEHLER,"Metallographic Practice and Hardness Testing for the Fastener Industry", BUEHLER LTD.,VOL. 26, NO. 1, 1990, Lake Bluff IL. USA.

9. Konig, W., Celi, I. and Noken, S."Stereolithography Process Technology" 3rd Int. European Conf. on Rapid Prototyping, Edited by Dickens, P. M., University of Nottingham , 6-7 July 94, pp 191-208, ISBN 0 9519759 2 7.

10. Parker, Sybil P."Optics Source Book", McGraw Hill, 1988, New York, ISBN 0 07 045506 6.

11. Crafer, R. C. & Oakley, P. J."Laser Processing in Manufacturing", Chapman & Hall, 1993, Cambridge, G.B, ISBN 0 412 41520 8.

12. Jacobine, A., et al., "New Photopolymer Resins for Stereolithography", 1st Int. European Conf. on Rapid Prototyping, Edited by Dickens, P. M., University of Nottingham, 6-7 July 1992, pp 163-181, ISBN 0 9519759 0 0.

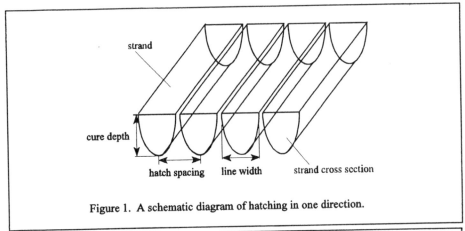

strand

cure depth

hatch spacing   line width   strand cross section

Figure 1. A schematic diagram of hatching in one direction.

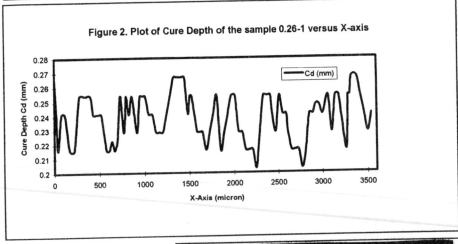

Figure 2. Plot of Cure Depth of the sample 0.26-1 versus X-axis

Cd (mm)

Cure Depth Cd (mm)

X-Axis (micron)

Figure 3 Cross section of sample 0.26-2 in X direction.

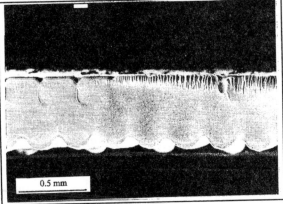

0.5 mm

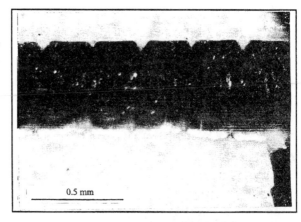

Figure 4 Cross section of sample 0.26-2, showing the partial staggered layers in Y direction.

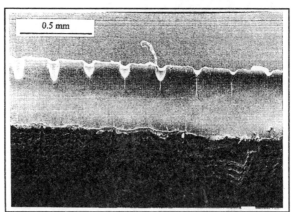

Figure 5 Cross section and underside of sample 0.26-2.

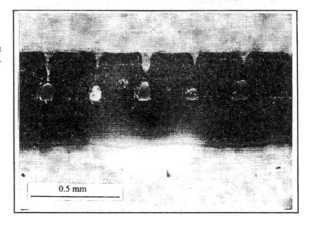

Figure 6 Cross section of sample 0.26-2, showing the vector cavities.

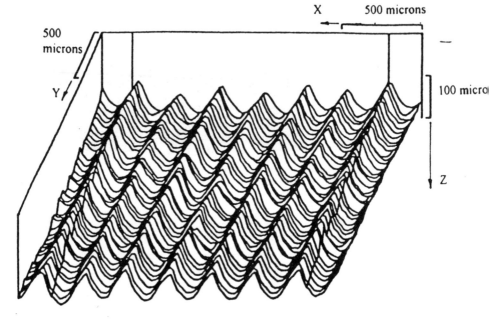

Figure 7 Underside of sample 0.26-2 by profilometry in X direction

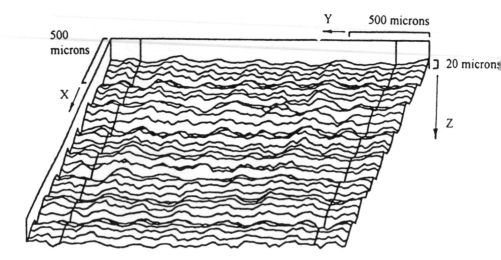

Figure 8 Underside of sample 0.26-2 by profilometry in Y direction

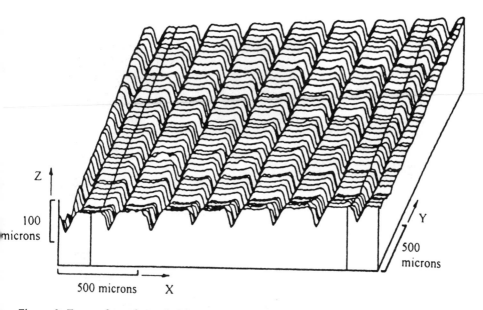

Figure 9 Top surface of sample 0.26-3 by profilometry in X direction

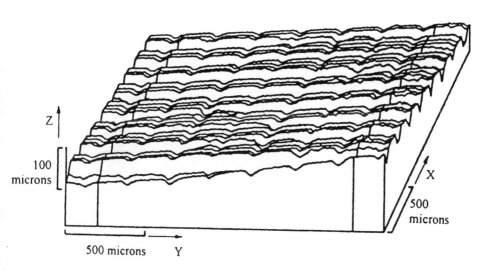

Figure 10 Top surface of sample 0.26-3 by profilometry in Y direction

Figure 11 Narrow hatch spacing of sample 0.22-2, avoids creation of local maximas, as a result a flat surface is obtained at the bottom and the top, It also shows incomplete retractions from borders.

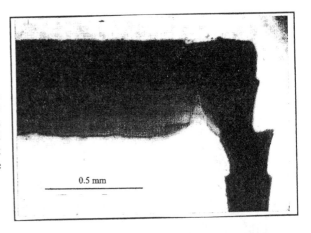

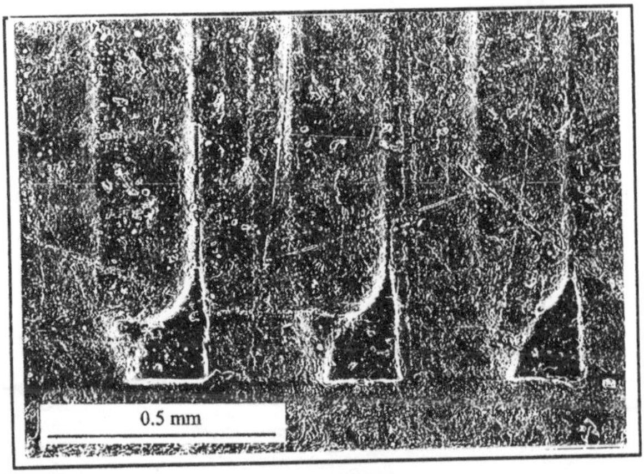

0.5 mm

Figure 12 Sample 0.24-2, showing incomplete retraction due to non-stop laser curing before a new strand begins.

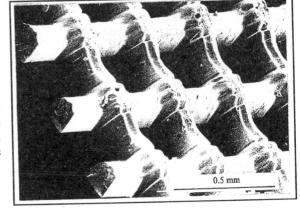

0.5 mm

Figure 13 Sample with 0.5mm hatch spacing showing the cross section of the cured photo-polymer, exibiting the difference between the first pass (horizontal) and the second pass (vertical) axis.

0.5 mm

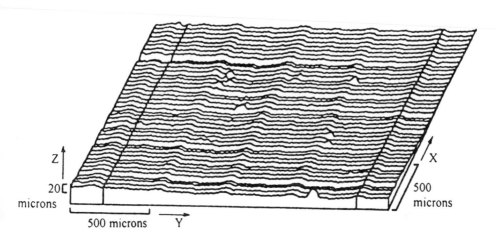

Figure 14 Top surface of sample 0.21-2 by profilometry in Y direction

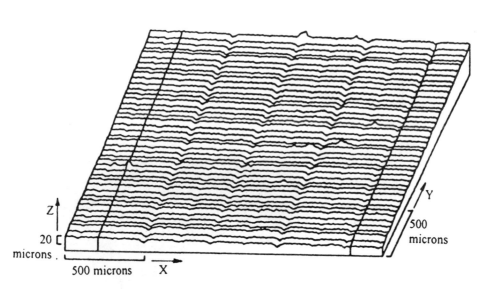

Figure 15 Top surface of sample 0.21-2 by profilometry in X direction

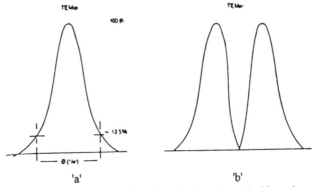

Figure 16 Schematic intensity profiles of an ideal and a typical laser beam.

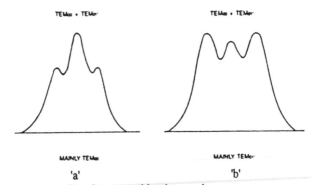

Figure 17 Intensity profile of two combination modes.

Figure 18 Double layer cross section of sample 0.26-2 showing, first the difference in layer thicknesses, second extended 'overcure' in Y direction to the support walls.

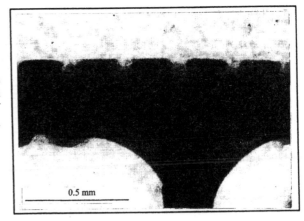

oduction of metal parts featuring heavy sections
ing 3D welding

SPENCER and P M DICKENS
ersity of Nottingham, UK

SYNOPSIS

3D welding offers the possibility of the direct production of homogeneous metal parts from CAD data. Experimentation has shown that the process is applicable to the fabrication of automobile castings.

It has been found that where sections greater than 6mm are specified, single beads are not practical. Adaptation of spiral overlay welding using multiple beads has enabled the production of parts wider than normally possible from a single bead.

Results are presented to compare the mechanical properties of two different test part geometries, one built horizontally and a similar, vertical structure built using the same technique. Tests include tensile testing and measurements of residual stresses.

1. INTRODUCTION

Rapid Prototyping technology has quickly established itself as an invaluable tool in the quest to reduce the lead time of new products from concept to product. New techniques to achieve this are constantly emerging. Amongst these is the application of 3D welding as a means of producing homogeneous, near net shape, prototype metal parts (1). Results suggest that the technique will be particularly useful in the prototype manufacture of medium to large automobile castings (2). Following exhaustive welding trials, it is now possible to produce single bead width walls from

3 to 6mm wide. Through experimentation it has been found that, when using 1.0mm diameter welding wire, it is practical to build single bead width walls no greater than 6mm wide. Above this heat inputs are excessive and bead profiles lost due to insufficient cooling rates. Thus, to build sections greater than 6mm we require multiple bead widths. It has also been noticed that certain part geometries have contained high levels of residual stress.

2. MULTIPLE BEAD WIDTH WALLS

When attempting to build parts in the horizontal plane from multiple welds (placing one beside the other), it was found to be extremely difficult to achieve penetration to the substrate and the neighbouring bead (Fig. 1). It was found that complete penetration could be achieved by increasing the current. However, this resulted in the loss of bead profile. Angling the welding torch into the intersection of the substrate and the previous weld was not considered a practical option to achieve penetration. This technique is very much an operator based skill and would be difficult to translate to the off line programming of complex 3D welded parts. Also the resulting bead profile would be unpredictable. Because the 3D welding process is to be used for the production of prototype near net shape parts it is necessary that we achieve maximum possible accuracy by producing predictable welds. Thus an alternative technique for producing multiple bead width structures was required.

0.5mm

Fig. 1. Lack of fusion when producing beads side by side indicated using die penetration.

3. DOUBLE SPIRAL OVERLAY WELDING

A review of previous work showed that a technique known as 'Double Spiral Overlay Welding' had been successfully applied for cladding stainless steel to the outside of mild steel bars (3). A single weld bead is deposited around the substrate bar in a manner similar to single point screw cutting but with material being added rather than removed. The pitch is such that the weld bead does not overlap itself. Thus the weld has a pitch, a crest and a root (Fig.2). Welding is repeated at the same pitch but at half a pitch increment such that the root gap between the weld is filled resulting in a continuous cladding (Fig.3). When performing the secondary (root filling) operation, certain welding parameters are altered so that fusion and penetration of the primary (boundary) bead is achieved by the fill bead.

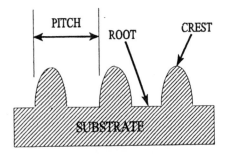

Fig.2. Definition of the features of a double spiral.

Fig.3. Complete cladding.

4. ROOT-FILL WELDING

The principles of Double spiral overlay welding has been adapted to the building of free standing, 3D welded structures. However, rather than welding around a substrate bar, weld beads that represent the extreme surfaces of a solid section(i.e. the inside and outside surface) are deposited on a surface. The area between them is sub-divided by further similar beads such that the spacing between the welds lies within a known range (Fig.2). The root is then filled with weld, the settings being such that fusion and penetration of the boundary beads and substrate is achieved, whilst the correct level of filling is attained. To avoid over complication and to ease off line programming we chose to use one (repeatable), stable weld to produce the root boundaries. This weld bead was selected from a data base of welding conditions compiled from the results of the trials used to investigate the effects of multiple welding parameters. The selected weld is of narrow width (3mm) and exhibits a high height to width ratio, which is beneficial when building 3D welded structures. Experiments have shown that the selected bead is suitable for additive layer welding. The weld is achieved at a relatively slow torch velocity thus allowing greater control of torch motion. At present the root-fill welding settings are selected against a reference of root widths. However, it is envisaged that in the future root - filling may be carried out automatically, perhaps using some form of through-arc sensing to ensure adequate root filling (4).

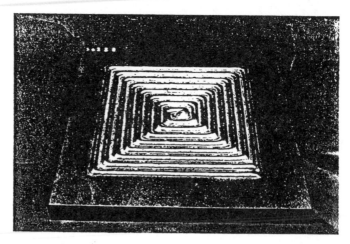

Fig.4. Area between boundaries subdivided into smaller regions for homogeneous filling.

The root fill method takes advantage of the ability to alter the welding parameter to produce different weld cooling conditions. The heat input during root filling may be influenced by the welding parameters to affect the cooling rate (5). Because the weld bead is bounded, material flow is constrained and thus higher heat inputs are possible without effecting the definition of the part. Future work will be carried out to discover how the welding conditions can be controlled so that parts can be produced with the minimum of residual stress which are known to reside within the 3D welded structure, caused by thermal stresses resulting from high heat inputs. This is clearly a problem and is clearly indicated as distortion of the substrate upon which the structures are produced.

5. THE WELDED TEST PARTS

Two similar, but technically very different, test parts have been created using the 3D welding technique. These were designed specifically to identify production difficulties at an early stage. The first is a solid, horizontal block measuring 100x100mm x10mm high, the second being a solid vertical block measuring 100x10mm x100mm high (Fig 5.). Both have been produced using the 'root fill' technique, the vertical part comprising of three beads per layer (two boundaries and one in fill) and the horizontal part being composed of thirty three beads per layer. Both parts were produced on 150x150 x 12mm thick stress free mild steel base plates which were restrained by clamping at each corner to a 50mm thick mild steel base prior to welding.

The material used in the production of both types of test part was NOVOFIL 70 SG2, a copper coated mild steel, 1mm diameter wire (C=0.08, Mn=1.5, Si=0.90, P=0.01, S=0.01, Cu=0.20). This material complies to the following standards; BS2901 A18, AWS E70 S-6, and DIN 8559 SG2. The manufacturer claims the following mechanical properties for this material; Tensile strength of over 540 N/mm^2, Yield strength of over 450N/mm^2 and an impact strength of 120 J. Because the strength of the material is high and its weldability good, it was deemed as being suitable to attempt to build 3D welded parts from.

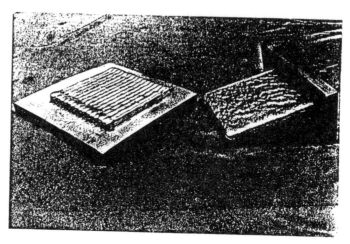

Fig. 5. Vertical test part (right) and Horizontal test part (left).

6. RESULTS

6.1 Tensile testing

The test parts were removed from their bases before being machined into test specimens. The test pieces being produced in accordance with, and tested in compliance with ISO 6892, Metallic materials - Tensile testing. For results of tensile tests please refer to Table 1.

6.1.2. Vertical Slab STP

Tests show that there are wide variations in the strength of the vertical slab test parts, from the bottom to the top of the slabs, the top being on average 8.6% weaker in terms of Ultimate Tensile Strength (UTS). Samples taken from the top recorded an average UTS of 517 N/mm^2 whereas those from the base showed an average UTS of 476 N/mm^2. Yield strength and stiffness (percentage elongation to failure) values were also very much lower for samples taken from the top of the wall. Samples cut from the centre portion of the plates, across the weld direction, showed very little variation in strength, all but one (STP-V-2-YL = 480 N/mm^2 UTS) recording a UTS of in excess of 490 N/mm^2.

Tensile test results have shown that, in some instances, the welded structures have similar physical properties to the drawn (work hardened) welding wire from which they have been produced. Tests indicated that whilst parts produced by 3D welding are reasonably isotopic, the physical properties of the parts are greatly influenced by inclusions and voids between the beads which are manifested when testing perpendicular to the direction of weld deposit (Table 1).

Table 1. Tensile test results of Horizontal and Vertical structures.

	MATERIAL	PLANE	SAMPLE NO.	% ELONGATION TO FAILURE	YIELD STRESS N/mm2	UTS N/mm2
Vertical slab						
	M/S	0	STP-V-1-XT	38	326	475
Top	M/S	0	STP-V-2-XT	35	276	479
	M/S	0	STP-V-3-XT	37	339	474
	M/S	0	STP-V-1-XB	22	427	512
Base	M/S	0	STP-V-2-XB	29	503	544
	M/S	0	STP-V-3-XB	22	427	496
	M/S	90	STP-V-1-YM	33	415	495
Centre	M/S	90	STP-V-2-YM	35	396	492
	M/S	90	STP-V-3-YM	34	389	492
	M/S	90	STP-V-1-YR	33	402	490
Right	M/S	90	STP-V-2-YR	37	380	497
	M/S	90	STP-V-3-YR	33	377	490
	M/S	90	STP-V-1-YL	35	377	494
Left	M/S	90	STP-V-2-YL	32	352	480
	M/S	90	STP-V-3-YL	35	415	491
Horizontal slab						
	M/S	0	SPAS-I5-B	16	459	565
	M/S	0	SPAS-15-E	18	459	539
	M/S	90	SPAS-15-A	4.5	350	384
	M/S	90	SPAS-15-C	4	344	366
	M/S	90	SPAS-15-D	6	459	540

6.1.1. Horizontal slab

Test on the Horizontal STP's show that whilst the strength of the material, in the direction of the weld, is good, inclusions/ voids have greatly reduced the strength of parts when tested across the direction of the weld. However in parts that have not failed prematurely, strength and stiffness have been found to be good (459 N/mm^2), although below the recorded strength of the material tested in the weld direction (539 - 565 N/mm^2 UTS).

6.2. Micrograph analysis ·

We have been able to successfully demonstrate the usefulness of Root - fill welding by building multiple bead width structures in both the horizontal and vertical plane. Macrograph analysis of sections show that the structural integrity of horizontal, multiple bead structures produced by this technique.is good (Fig. 6). Fusion between the beads has been achieved and penetration has occurred. The macrograph shows that the upper surface of the horizontal structure, having rapidly cooled, exhibits a course, hard, martensite structure. Re heating due to overwelding, and slower cooling of the underling layers has resulted in a finer ferrite / pearlite grain structure in the main body of the structure.

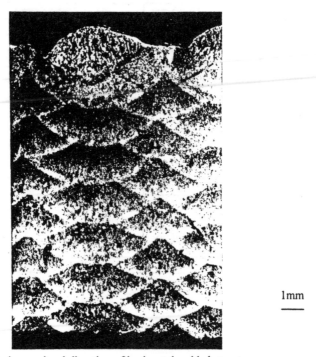

1mm

Fig. 6. Sectional view on bead direction of horizontal welded structure.

Similar results to those found in the horizontal structure were obtained with the vertical structure. Fig.7 shows a cross section of three fused weld beads which compose the vertical test part, the section has been taken from the centre of the structure. As with the horizontal test part, fusion between the weld beads has been achieved. Also variations in the crystalline structure can be observed throughout the part. The occurrence of these different crystalline structures may explain the differences in the physical properties of test samples cut from various localities of the structure.

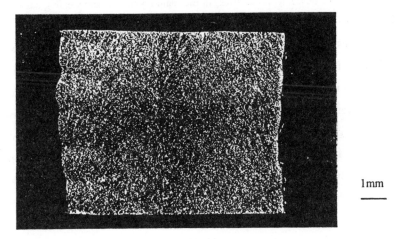

1mm

Fig. 7. Sectional view on bead direction of vertical welded structure.

7. RESIDUAL STRESS

In common with many other processes where material is subject to phase change (eg. liquid to solid transformation) shrinkage of the material occurs and is influential in the formation of residual stress in 3D welded parts.

The levels of residual stress present in all welded structures is dependant upon material strength and thermal properties and the degree of restraint that prevent free contraction of the solidifying material and adjoining material. Mechanisms of constraint can be of the following nature,-

longitudinal

transverse

substratum

Residual stresses are caused as a result of thermal stresses induced by the welding process, this causes the material about the weld to expand. Because the material strength is reduced due to increased temperature, yielding and permanent plastic deformation occurs. Once the heat is removed (90% by conduction) the weld bead and surrounding heated area contract.

The weld and the material in the immediate vicinity of the weld have been subjected to greater heating and has thus undergone greater thermal expansion than regions more remote from the heat source, hence contraction in this region is greater once heat is removed. As shrinkage occurs so tensile stresses are exerted in regions about the weld bead. Further from the weld these tensile forces tend to be compressive stress in the region where plastic yielding has occurred, caused by thermal stress and a softening of the material (6).

Structures produced by 3D welding at the University of Nottingham have exhibited various levels of residual stress, depending upon the geometry of the part produced. Measurements of residual stresses using the hole drilling method have shown that high levels of residual stress ranging from 680 to 840 N/mm^2 are included in the horizontal block test parts which have a large bead to base interface, compared to a maximum of only 94 N/mm^2 at the centre of a 100 x 100 mm single bead width vertical wall (7). Stresses present in the horizontal block can be seen by distortion of the base plate to be greatest in the direction of the weld. Because no preheating was employed there has been a large temperature gradient between the deposited weld bead and the base plate when welding began. This combined with the mass of heat sink available results in rapid cooling of the weld and surrounding material preventing the material from plastically deforming at reduced applied stresses.

7. CONCLUSION

The adapted double spiral or root fill technique has proved valuable in producing parts where wall thickness greater than 6mm are required. The technique allows building of walls upwards of 6mm wide, predictably and accurately. The parts exhibit good structural integrity and micrograph analysis shows that there is good joining between the beads. This is reinforced by the results of the tensile tests which show the physical properties of the welded structures approach that of the origin material. Further work will need to be conducted to find methods by which inclusions and voids can be eliminated. Investigations will also need to be made regarding the use of sensing to control root filling between the boundary beads. The nature and cause of residual stresses in 3D welded parts needs to be further investigated and studies conducted to find ways of minimising them.

References

1. McAninch, M.D. and Conrardy, C.C. Shape melting-a unique near-net shape manufacturing process, Welding Review International, 1991, Vol 10, pp 33 - 40.

2. Dickens, P.M.; Cobb, C; Gibson, I. and Pridham, M.S. Rapid Prototyping using 3D welding, Journal of Design and manufacturing, 1993, Vol 1, pp 39-44.

3. Kalligarakis, K. and Mellor B.G. Double spiral overlay welding- an alternative to single and multilayer techniques. Welding and metal fabrication, 1992 Vol 60, pp 277-280.

4. Bates, B.E and Hardt, D.E. A Real-Time Calibrated Thermal Model for Closed-Loop Weld Bead Geometry Control. Journal of Dynamic Systems, Measurement, and Control, 1985, Vol 107, pp 25 - 33.

5. Million, K; Datta, R. and Zimmermann, H. Method and manufacture of a design piece by form creating deposition welding and a design piece manufactured by this method. European Patent Application 85104225.9

6. Lancaster, J.F. Metallurgy of Welding, 1994, (Chapman and Hall), pp 159 - 162.

7. Report for Promet. EU project, Brite-Euram BE 5443.

Improvements in the surface finish of stereolithography models for manufacturing applications

E REEVES and **R C COBB**
University of Nottingham, UK

1 ABSTRACT

StereoLithography (SL) can produce accurate Rapid Prototype models suitable for a wide range of applications. These can include aids to design verification, form and fit, finite element analysis (FEA) models and patterns for short run production tooling. A major barrier to overcome before SL can be used more widely is the excessive time required to obtain an acceptable surface roughness, while maintaining part accuracy.

This paper details a number of factors in the SL process which have been associated with poor surface finish, each resulting in a cumulative effect on the overall part roughness. A methodology for surface finishing using additive processes, abrasive finishing and chemical etching is discussed.

Initial findings of the work show that additive coatings can improve surface geometry and roughness by up to 50% however, a combination of additive and abrasive techniques can result in a reduction in surface deviation of up to 95% on complex surfaces.

2 INTRODUCTION

Current surface finishing techniques applied to StereoLithography models are highly selective, and often neglect fine detail and internal features. The processes adopted are also labour intensive, time consuming and not cost effective.

Through the introduction of new resins and build styles machine manufacturers have increased part accuracy and surface finish almost six fold since 1989. The introduction of epoxy resin and the ACES™ build style has resulted in parts with roughness average (Ra) values of less than 0.28 µm on up-facing planes, this being a significant reduction from the early acrylic components with roughness values of 16µm Ra on equivalent surfaces (1). With the introduction of the new Quick-cast™ 1.1 build style, it is possible to achieve surface roughness values of 0.25 - 0.5µm Ra on sacrificial patterns used for investment casting (2). Quick-cast™ 1.1 uses triple up facing layers to reduce deviation by applying additional layers of material to the part, covering any deviation caused by previous layers.

Research into these new build styles has however addressed only the problems associated with vertical and horizontal planes. Analysis of the overall surface roughness of SL parts suggests that the surface roughness on angled planes require greater research, irrespective of either build style or material. It can be seen in Figure 1 that although both vertical and horizontal planes produce low roughness values, angled planes generate highly stepped profiles.

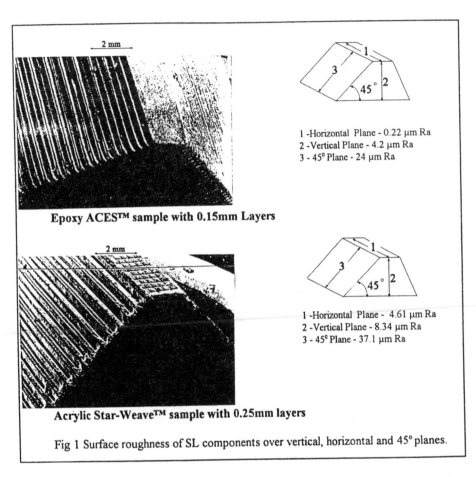

Epoxy ACES™ sample with 0.15mm Layers

1 - Horizontal Plane - 0.22 µm Ra
2 - Vertical Plane - 4.2 µm Ra
3 - 45° Plane - 24 µm Ra

Acrylic Star-Weave™ sample with 0.25mm layers

1 - Horizontal Plane - 4.61 µm Ra
2 - Vertical Plane - 8.34 µm Ra
3 - 45° Plane - 37.1 µm Ra

Fig 1 Surface roughness of SL components over vertical, horizontal and 45° planes.

In many cases the problems of surface deviation are generated at the CAD stage such as approximation to curves and surface tessellation.

One such limitation is the StereoLithography Transfer (STL) format, used to transfer CAD geometry to the RP system from the design environment. The data is only an approximation of the CAD surface, and can result in loss to part definition through faceting of non-plane surfaces and features. A new transfer format has however been developed to eliminated the need for surface tessellation. The StereoLithography Contour (SLC) format transfers CAD representation as a series of extremely thin '2D' layers in much the same way layer manufactured object are created, the layers are then re-stacked following data transfer (3). Although SLC precludes the generation of tessellated parts, present CAD software only supports the tessellated STL format, generated models with faceted surfaces.

Work has also been undertaken to reduce surface roughness on complex planes and features, using part orientation software prior to slicing (4). The STL representation of the CAD data is oriented about its axis in such a way that layering will be reduced on planes perpendicular to the 'Z' axis, hence reducing 'stair stepping' . However, with this software the effects of trapped volumes, delamination, build time and working envelope must all be considered in addition to the positioning of support structures required for model stability (5). With many complex parts, software orientation can only reduce 'stair stepping' on a limited number of surfaces, and additional post process finishing is still required.

Given that all SL models at present are at best only stepped, and at worst badly faceted, research has been undertaken to establish faster more consistent finishing techniques, using conventional mass finishing technology. The Brite-Euram INSTANTCAM project (6) investigated a number of SL components using 'traditional' abrasive flow tumble peening and sandblasting equipment. Work at the University of Nottingham applied acrylic SL parts to a range of automated finishing equipment including barrel tumbling, vibratory finishing, ultrasonic abrasion and abrasive flow blasting (7). Some encouraging results were observed but many of the component were found to have excessive damage, with loss of material at both edges and corners (see Figure 2). In both cases, as with work undertaken at the GINTIC institute on SL abrasion (8), the abrasive systems employed were developed for the finishing of metallic components using harsh ceramic media. Using such media types on plastic components produces SL parts with excessive edge radiusing and surface damage.

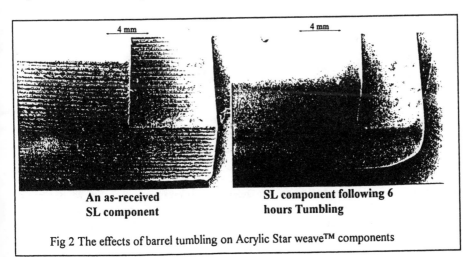

| An as-received SL component | SL component following 6 hours Tumbling |

Fig 2 The effects of barrel tumbling on Acrylic Star weave™ components

It is the intention of this paper to report on research work carried out within an EPSRC funded project 'Rapid Finishing and Tooling'. The paper will identify some of the factors involved within the StereoLithography process that are responsible for the surface deviation of models. The application of resin based coatings with some mechanical finishing is then discussed.

3 ANALYSIS OF SL SURFACE ROUGHNESS

The effects of layer thickness or 'step height' can be considered a major contribution to surface roughness, particularly for incline planes as shown in Figure 1. Hence, research at the University of Nottingham has been directed towards establishing the deviation in surface roughness of StereoLithography models. As shown in Figure 3 a mathematical representation of a surface can be made, and by using simple trigonometry a surface Ra value can be derived given the assumption that SL layer edges are perpendicular rather than slightly parabolic.

Using the surface roughness measurements of a range of SL models, a comparison can then be made between 'derived' and 'actual' roughness values for a range of angled plane surfaces. Figure 4 details both the mathematically derived surface roughness and the average Ra value of SL planes in 10 degree increments from vertical to horizontal.

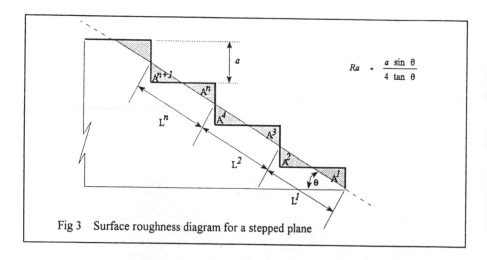

$$Ra = \frac{a \sin \theta}{4 \tan \theta}$$

Fig 3   Surface roughness diagram for a stepped plane

As it can be seen from Figure 4, the actual surface roughness measured across a range of surface planes give lower Ra values than mathematically predicted. The effects of the parabolic curing profile should in theory increase the surface roughness however, analysis suggests that excess resin not removed during part stripping results in a reduction in surface Ra.

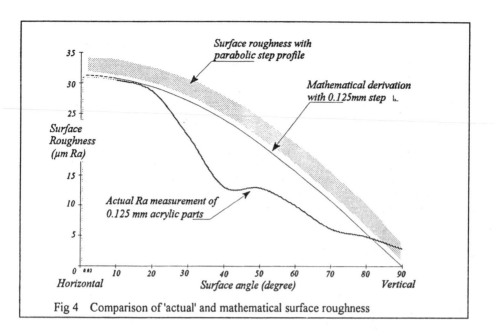

Fig 4   Comparison of 'actual' and mathematical surface roughness

4 RESEARCH METHODOLOGY

The presence of additional material as part of the build step can therefore be considered to provide SL parts with lower Ra values compared to those which are thoroughly cleaned prior to post curing (9). Hence, if additional material is added to the build step, the surface roughness of the part is likely to be reduced. The addition of coating materials to a range of standard geometry parts is one of three research areas currently under investigation within the research project . Material can be added to the build step in a liquid state by spraying, dipping or painting, or material can be removed from the part surface by abrasive machining or chemical etching with acids and solvents.

The basis of the research methodology is that, SL parts can be built with either the CAD geometry at the outer or inner edge of the build step, hence in order that the part be 'near net shaped', material must be either added to fill step cavities or it must be removed to reduce peaks.

5 PHOTOCURABLE RESIN COATINGS

From the comparisons of SL parts with the mathematically calculated surfaces the use of StereoLithography resin seems effective for reducing Ra values. Experiments involving SL resin as a coating were thus conducted on both acrylic and epoxy parts in both the cured and un-cured 'green' state (9). Parts were shown to have excessive loss of geometry, due to the high viscosity of the resin coatings. A thinning agent, Hexandiol-diacrylate was added to the resin and thinner coatings could be applied to parts without excessive loss to part geometry.

The thinned XB5149 acrylic resin seemed to show some potential although due to the diluent nature of the solution, cross-polymerization of the monomer chains were not possible, and the coating remained 'tacky' following UV post curing. Photocuring of the SL diluent was also performed in a nitrogen enriched atmosphere, resulting in cross polymerisation through the inhibition of oxygen. Although the coating resulted in an SL part with a lower surface roughness, loss to part geometry was evident with holes and recesses blocked.

In summary, the application of diluent SL resins has potential, particularly prior to post curing as this would minimize the number of manual operations in the construction of a model. At present however the processing time with solution preparation, application and nitrogen curing is labour intensive and difficult to justify. It is intended to continue the research into a suitable diluent, capable of producing a low viscosity solution with full photo-polymerisation.

6 EPOXY BASED RESIN COATINGS

The initial trials with SL resin diluents showed that resin coatings had potential for reducing the surface roughness provided that the coating thickness, which is a function of the initial viscosity, can be defined and controlled. Alternative coatings possessing good wetting and adhesion characteristics were therefore investigated.

Jaxacote 22.22 is a commercially available three part epoxy loaded primer coating used in the manufacture of glass fibre composites, as a filling agent for the surface of damaged gel coats. The coating contains both a filler and a low temperature exothermic agent, in addition to a thinning agent which can be added up to 75% by volume to reduce the coating viscosity. The coating is formulated to have excellent wetting and adhesion to epoxy substrates, resulting in a high sheen finish. It can also be abraded following curing, to give a sub-micron surface roughness.

A series of experiments were undertaken to apply both concentrated and thinned epoxy primer to acrylic and epoxy SL parts by painting, spraying and dipping, with the intention of reducing surface deviation yet retaining or improving part geometry (see Figure 5).

Initial findings showed that the coating in its concentrated state can reduce surface deviation on vertical and horizontal planes from 5.1μm and 2.1μm to 1.9μm and 0.9μm Ra respectively, but the coating viscosity led to significant loss to part geometry (see Figure 6).

Following investigation with the concentrated solution, a 50% thinned coating was applied to both acrylic and epoxy samples. At the lower viscosity it was necessary to apply multiple coatings in order to reduce surface roughness by significant levels. At the thinned viscosity multiple layers of the epoxy primer applied by both dipping and painting were found to result in surface reduction of up to 50% on stepped planes. After two layers applied by dipping vertical and horizontal planes were reduced from 5.1μm and 2.1μm to 1.2μm and 0.6μm Ra respectively. The effect of the epoxy primer as seen in Figure 7, was not only to reduce the Ra value of the component, but also it improved the geometry to nearer that of the intended CAD profile.

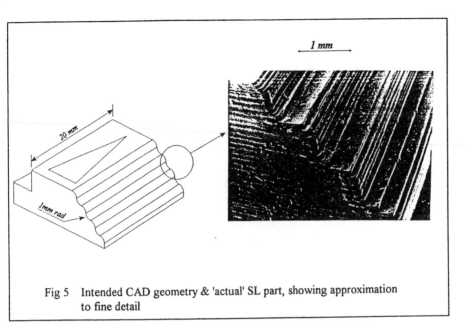

Fig 5   Intended CAD geometry & 'actual' SL part, showing approximation
to fine detail

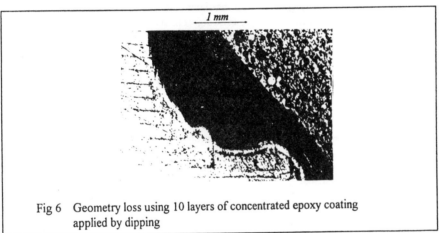

Fig 6   Geometry loss using 10 layers of concentrated epoxy coating
applied by dipping

Results have shown that thinned epoxy primer (Jaxacote) is capable of reducing surface roughness on parts of limited complexity with only minimal blocking of holes and features, although the time taken to build up sufficient coating thickness needs to be reduced.

1 mm

Fig 7   SL sample coated in two layers of 50% thinned epoxy primer
applied by dipping

7 DUAL PROCESSING

One method of reducing the lengthy 'manual' finishing time for SL models is by automated abrasive finishing. For 20 years thermosetting plastic object manufactured from polymers such as ABS, acrylics, polyvinyl butride and impact styrene have been 'mass finished' in automatic systems to remove excess material such as lugs, flash and machining marks (10). However, previous attempts to finish StereoLithography parts (6,7,8) have shown that conventional 'mass finishing systems' cause considerable damage, but in each case a 'ceramic finishing media' and systems intended for the abrasion of ferrous and non-ferrous metals have been used, this would appear to be too harsh for the polymer material.

The system used within the University of Nottingham is a barrel finishing or 'rumbling' machine run at approximately half conventional speed, with a hardwood lining to prevent part damage. The media used is a 'solution' of fine ceramic 'silica' powder (0.25μm mesh) bonded using a resin agent to a carrying media. Carrying media can take the form of either natural hardwood or polyester moulded preforms (11) ranging in size dependent on the application.

Both acrylic and epoxy standard geometry parts have been finished using this system, and results will be presented at a later stage. However, some parts were coated in a thick 150μm layer of concentrated epoxy and subjected to barrel finishing. Although previous experiments have shown that 50% thinned primer gives a better surface finish the concentrated solution was used to build a thick layer rapidly, as the coating is now used sacrificially to reduce surface deviation. The media is then used to remove excess material from 'outside' the build step, but due to the media geometry is prevented from abraded material within the step profile.

Test were carried out using eight epoxy Quick-cast™ aero-engine turbine blades, which were used to assess the effectiveness of the coating and finishing treatment on 'real' industrial components. Following coating, the 150μm layer of epoxy was cured at ambient temperature for 4 hours. The coated parts were then barrel finished in the system for 45 minutes per blade (6 hours total process time), following which an analysis of both geometry (see Figure 8) and the surface roughness using a Taly-surf were undertaken (see Figure 9).

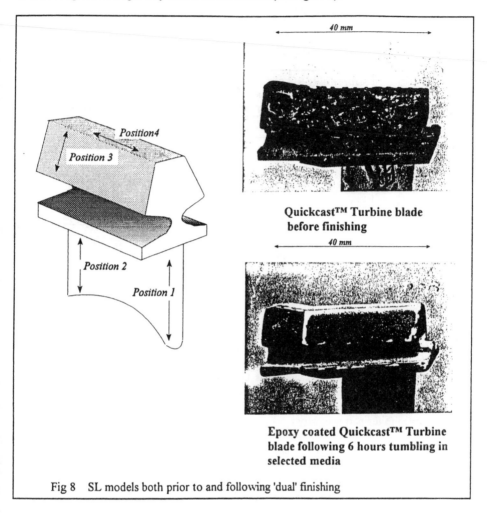

Quickcast™ Turbine blade before finishing

Epoxy coated Quickcast™ Turbine blade following 6 hours tumbling in selected media

Fig 8   SL models both prior to and following 'dual' finishing

Analysis of the surface has shown that by 'dual' processing it is possible to reduce surface deviation on a complex component by up to 95% over the total surface area, with minimal damage to the part geometry. Trials with coated parts showed a marked reduction in surface deviation however there was a general loss to the part geometry and definition without additional tumbling. Abrasion of un-coated components showed little improvement to the overall surface roughness, this being attributed to the harness of the cured SL resin being greater than that of the epoxy coating. The abrasive system can therefore be considered suitable for the abrasion of the epoxy coating without damage to the SL substrate. Analysis of parts was not undertaken prior to six hours however further research is intended to establish the minimum time period for post process abrasion resulting in the greatest surface reduction.

POSITION	Surface roughness or supplied blade (μm Ra)	Surface roughness of blade following tumbling (μm Ra)	Surface roughness of 150μm coated blade (μm Ra)	Surface of 'dual' processed blade (μm Ra)
1 - Leading edge	6.38	5.5	2.3	1.8
2 - Trailing edge	5.36	5.2	1.9	1.25
3 - Across fir tree root	11.6	11.1	3.6	1.6
4 - Along fir tree root	25.7	24.8	3.4	1.29

Fig 9   Surface roughness of turbine blades following processing

8 CONCLUSIONS

In conclusion it can be stated that, a number of positive directions for future research have been established. The use of thinned SL resin as an additional coating, will if successful, result in near transparent SL models requiring no other post process finishing, yet retaining their intended geometry. The epoxy primer (Jaxacote) has been shown to be an excellent coating for SL components, either as a base for other coatings or as a method of surface finishing. With the application of thinned primer by dipping it is possible to reduce surface deviation rapidly with minimal change to the part geometry.

'Dual' additive and abrasive finishing has been seen to reduce surface deviation on complex models by up to 95%, with minimal change to the part geometry. It is intended through further research to optimize the abrasive media and system to result in a fast, non-selective finishing system capable of reducing surface deviation to a better quality and more economically than current finishing techniques.

9 ACKNOWLEDGEMENTS

The authors would like to thank the following bodies for there contribution to this research:-
EPSRC, Rolls Royce (Aerospace group), Britains Petite, Sonneborn & Reick, Ciba GEIGY, Lea Manufacturing.

10   REFERENCES

1.   JACOBS, P.F. JOUNI, P. BEDAL, B. Surface Quality, INSIGHT, 3D systems, 1994.

2.   JACOBS, P.F. Quickcast™ 1.1 & Rapid Tooling, 4th European conference on Rapid Prototyping and Manufacturing, Belgirate, ITALY, June 13th, 14th & 15th, 1995 pp 1 - 27, (The University of Nottingham).

3.   ASHLEY, S. New Rapid Prototyping file format, Engineering, April - May, 1995.

4.   FRANK, D. Preferred direction of build for Rapid Prototyping processes, 5th International conference on Rapid Prototyping, Dayton, USA, June 12th, 13th, 14th & 15th, 1994, pp191 - 200 (RPDL University of Dayton).

5.   ALLEN, S. On the computation of part orientation using support structures in layered manufacturing, Solid Freeform Fabrication symposium, The University of Texas at Austin, August 8th, 9th & 10th, 1994, pp259 - 269 (The University of Texas).

6.   INSTITUTO SUPERIOR TÉCNICO. INSTANTCAM - Work Area 3 - Final report BriteEuram Project No. BE-3527-89, Report WA3, Instituto Superior Técnico, PORTUGAL, 1992.

7.   SPENCER, J. DICKENS, P.M. COBB, R.C. Surface finishing techniques for Rapid Prototyping - Technical paper PE93-169, Rapid Prototyping Conference Dearborn, Michigan, USA, May 11th, 12th & 13th, 1993, (Society of Manufacturing Engineers).

8.   BOON, L.E. CHUA C.K. Case studies of Jewellery Prototyping using the SLA, Solid Free Form Manufacturing, 1st International user congress, Technical University of Dresden, GERMANY, October 28th, 29th & 30th, 1993, pp 1 - 18.

9.   REEVES, P.E. COBB, R.C. The finishing of StereoLithography models using resin based coatings, Solid Freeform Fabrication symposium, The University of Texas at Austin, USA, August 7th, 8th & 9th, 1995, (The University of Texas).

10.   DAVIDSON, D.A. Mass finishing of plastics, Product finishing, July 1984, pp 44 - 47

11.   DAVIDSON, D.A. Current developments in dry process mass finishing, Finishers Management, Vol 33, No 7, September 1988, pp 43 - 46

apid prototyping and tooling with a high powered ser

, THOMSON and M S PRIDHAM
versity of Dundee, UK

SYNOPSIS

Commercial rapid prototyping techniques are now well developed, however one of their main shortfalls is that they, in the main, produce parts in non metallic materials.

Two potential routes to circumvent this difficulty are presented in this paper. The first is a variant of the LOM technique, using sheet metal rather than paper or plastic and building tooling as opposed to parts. The second technique is a direct process for forming sheet based components without the need for tooling. This technique relies on controlled laser energy to deform the sheet in a precise and repeatable manner.

1. INTRODUCTION

Now that a number of rapid prototpying techniques have become well established it is becoming evident that one process is unlikely to be able to fulfil all the demands and requirements that are being made on these technologies, beyond the basic ability to produce a prototype model.

In addition to fit, form and aesthetic evaluation, functional testing can demand material properties and processing close to those encountered in production situations. One approach to this problem has been to use say an SLA model as an investment casting pattern(1). This works successfully but clearly dilutes the advantage of a single step production route such as stereolithography.

There are considerable moves toward using metallic materials directly in rapid prototyping techniques such as those going on in laser sintering and laser cladding (2) as well as the metal based lamination and laser forming techniques presented in this paper.

In addition to the rapid realisation of prototype parts, rapid prototyping is being exploited as a route to the low lead time production of development and even production tooling. Techniques which can eliminate or reduce expense and delay in producing such tooling offer many advantages in the design and development of a product and, as indicated above, even in production itself. Producing prototype parts by any means which involves production, and in most cases alternation and modification, of tools or moulds is a lengthy and costly procedure.

The stereolithography RP route is addressing this issue with developments such as Quickcast (1) and Quickcast 1.1 (3). An alternative is to produce the tooling directly in metal by a one stage process and this is being done with the development of laminated metal tooling which, in addition to producing tooling quickly, can offer the added advantages of allowing the tool to be easily altered and modified or re-configured.

The direct production of sheet metal prototypes and parts is possible via the laser forming route. This technology has been the subject of considerable research (4) (5) but has yet to be fully developed and exploited.

2. TOOLING BY METAL SHEET LAMINATION
Laminated object manufacturing (LOM) is established as a successful RP technique, constructing parts using layers of paper or plastic.

Using a high powered laser machining centre, it is possible to carry out the same type of cutting operation using sheet metals such as steels. The problem arises in this situation in permanently joining the laminations in a satisfactory manner. A number of possibilities exist including welding, soldering or brazing, adhesive bonding or mechanical fastening. Each may be possible in particular circumstances but they are all limited in other fields of application. Adhesives are, in the main, easy to use and convenient and have been used to construct simple trial layered components (see figure 1), but lack the mechanical strength of the laminate material itself and, therefore, limit the stresses the part may be exposed to particularly in the plane of the laminations. In some instances, solders or brazes are more suitable, however the same arguments largely apply. Welding produces high strength joints but the effect of heat and distortion can be unacceptable particularly in thin walled components it is however possible, and successful, in some applications. Mechanical fastenings can be used where there is a large area of laminate to bolt or clamp through but in a finished component the fastenings themselves are likely to be at least obtrusive and at worst totally unacceptable. If however, the laminating technique is used to produce, for example, a mould cavity (6) rather than a finished part then the aesthetic issues associated with the mechanical joining technique can largely be eliminated. For example, large peripheral areas or flanges can be incorporated into the design to allow area for adequate mechanical fastening without interfering with the working area of the tool, see figure 2. In this figure a small four part (upper and lower sections and two inserts) mould, produced at the University of Dundee, is shown. The seventy or so laminae required were laser cut from 1 mm thick mild steel, each with its individual profile determined by its place in the tool, the fixing holes are also produced at this stage allowing easy assembly and alignment. In this particular case each half of the mould is assembled separately and the two halves are then bolted together. This tool has been used to produce wax mouldings for evaluation purposes. The wax was chosen for a number of reasons including its ease of melting and reuse, but more importantly because it faithfully replicates the surfaces against which it is cast giving a good indication of the surface finish obtainable. It should be noted that many materials when cast or moulded into a cavity of this type are likely to have a surface finish which is superior to that of the cavity. Clearly this can be an advantage in proudcution but makes evaluation of surfacette finish which may be possible difficult.

One of the issues in cavity surface finish is the 'stepping' which is unavoidable in any geometry other than vertical walls. In the example shown in figure 2, this is most severe in the regions approaching the poles of the truncated sphere as compared to the equatorial regions, it is also clearly influenced by the sheet thickness and the size of the features within the mould cavity. An ongoing programme, involving both additive and subtractive techniques, to address the surface finish issue is underway. An important factor in this work is that any finishing operation should ideally not effect the ability to disassemble, alter and reassemble the tool.

3. LASER FORMING

This process is able to induce precision, non contact deformation in sheet material by the application of a laser beam. The beam tracks over the surface of the material and heats the thin band of material directly beneath it. As the beam is moved on, the material behind it cools in either a forced or natural manner.

The forces set up by the heating, partial melting, resolidification and cooling of a very localised partion of material, in an otherwise unaffected mass, induce deformation. The degree and rate of deformation depend on such parameters as, tracking speed, laser power, degree of focus of the beam, material type and sheet thickness but a bend of one to two degrees on a single pass is typical. Figure 3 shows the effect of number of passes and tracking speed on the bend induced in 2mm thick stainless steel. By repeatedly tracking over the same line in the sheet the degree of bending increases, this can proceed so long as the beam can impinge directly along the fold line. As 90^0 is reached some reorientation of either the beam or the component is necessary to allow deformation to continue.

Where a gradual bend of a given radius of curvature is required this can be achieved by offsetting the track of the beam between passes. The component shown in figure 4 was made by a combination of sharp and gradual fold operations on a laser cut blank. Where more complex deformations are required, such as in raising or bending flanges, non uniform laser power may be required, that is to say the energy input must be varied along the bend line. This is necessary to take account of the fact that material close to a free surface for example, will be less constrained than that remote from it. To assist in these more complex geometries a finite element package is being using to help model the laser energy requirements to achieve the necessary deformation.

In the normal mode of operation the material will bend toward the beam, however, by altering the laser penetration and degree of focus it is possible to make the material bend away from the beam, whilst this can be useful, in many cases it is however more straightforward to simply reorient the part being formed this will be an important consideration in building a dedicated laser forming system, possibly as an element in a laser manufacturing facility also incorporating laser cutting and laser welding.

4. CONCLUSIONS

Laminated metal tooling is an area of RP which is now receiving a considerable amount of attention and interest. It offers a quick, low cost, versatile method of manufacturing development, or even production, tooling capable of producing relatively large numbers of components. Like all other rapid prototyping and tooling techniques it will find its own application niche. How large this is would seem to depend significantly on the ability to satisfy surface finish requirements.

As an exploitable process, laser forming is still in its infancy but with continued development, and work on the production of more complex geometries, together with systemisation of the process there is no reason sheet metal products can not be produced directly by this tool-less process.

ACKNOWLEDGEMENTS

The authors would like to express their gratitude to Mr Lyall Mitchell for his technical assistance, and to the Department of Applied Physics and Electronic & Mechanical Engineering for the provision of laboratory facilities.

REFERENCES

1. DENTON K. R., JACOBS, R. P., Qucikcast and Rapid Tooling, 3rd European Conference on Rapid Prototyping and Manufacturing, pp 53-72 University of Nottingham 6th-7th July 1994.

2. HAFERLKAMP, H., BACH, W., GERKEN, J., MARQUERING, M., Rapid Manufacturing, Direct Production of Metal Parts with Laser Radiation, 4th European Conference on Rapid Prototyping and Manufacture pp 123 - 136 Lake Maggiore, Belgirate, Italy June 13th - 15th 1995

3. JACOBS, P.F., Quickcast 1.1 and Rapid Tooling, 4th European Conference on Rapid Prototyping and Manufacture pp 1 - 26 Lake Maggiore, Belgirate, Italy, June 13 - 15 1995

4. FRACKIEWICH, H., Laser Metal Forming Technology Fabtech International '93, Chicago, Illinois, October 1993 pp 733 - 747

5. GEIGER, M., VOLLERSTEN, F., DEINZER, G., Flexible Straightening of Car Body Shells by Laser Forming Proceeding of the Sheetmetal and Stamping Symposium 1993 pp 37 - 44

6. GLOZER, G.R., BREVICK, J.R., Laminate tooling for Injection Moulding Proceedings of the Institute of Mechanical Engineers Part B Journal of Engineering Manufacture 1993 Vol. 207 pp 9 - 14

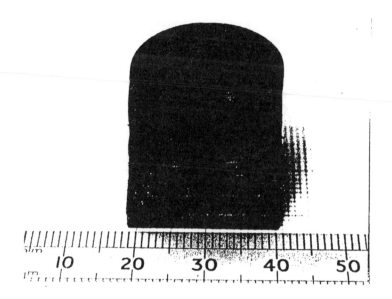

Figure 1 Metallic Laminated Part with Layers Adhesively Bonded

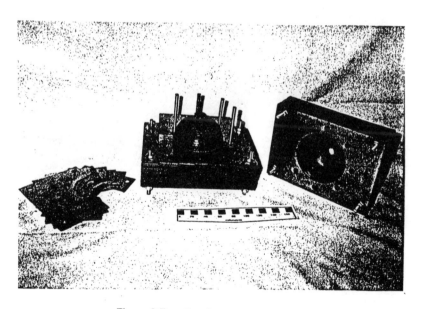

Figure 2 Four Part Laminated Tool

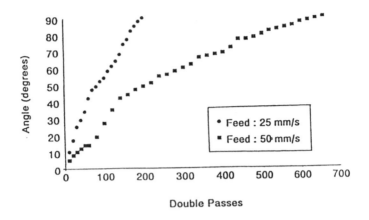

Figure 3 Effect of Number of Passes and Tracking Speed on Induced Bend

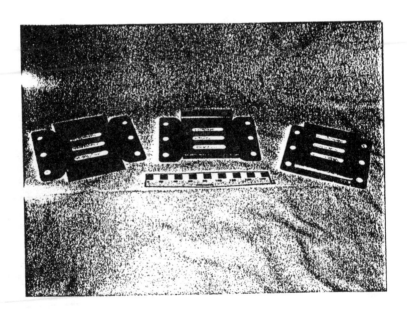

Figure 4 Laser Formed Cover Plate

investigation into the suitability of high speed
ective jet electrodeposition for rapid tooling

JOCKING
Marconi Materials, UK
OVER and **G BENNETT**
Nottinghamshire College of Higher Education, UK

SYNOPSIS

An investigation into the suitability of High Speed Selective Jet Electrodeposition for creating Rapid Tooling has been carried out. The surface morphology, rate of deposition and deposit profile have been investigated.

Experiments have been carried out to selectively plate nickel onto a steel substrate for the purpose of rapid presstool manufacture. The initial results of the characteristics of such deposits are reported. Gold has also been selectively deposited to investigate the limits of the apparatus used, and the overall feasibility of the technique.

. INTRODUCTION

Rapid prototyping is a fast growing activity within industry. It can offer incredible time-to-market savings with all the financial benefits involved. However, many companies are reluctant to buy a rapid prototyping machine, preferring to use bureaux. This can be put down to three reasons.

1) The price of the systems sold by the market leaders (SLS produced by EOS GmbH and the DTM Corporation, SLA produced by 3D Systems Inc. and EOS GmbH) are measured in the hundreds of thousands of dollars. Moreover, the costs of maintaining these machines are high. Additionally, those machines using exotic materials, e.g. photosensitive resins, further raise costs.

It is fair to say that machines that do not use lasers and the associated optical equipment e.g. FDM (Stratasys Inc.) and BPM tend to be cheaper.

2) The range of materials that Rapid Prototyping machines can use is limited. Although new materials are being developed continuously, the only commercially available systems which can produce metal parts directly is the EOS SLS machine (EOSINT), with the DTM Corporation reportedly on the verge of bringing a metal SLS machine onto the market. Metallic prototypes have many obvious material advantages and present the possibility of Rapid Tooling.

It should be noted that metal parts are achievable through multi-step processes e.g. investment casting and electroforming, and significant time savings are still available over conventional manufacturing methods.

3) Dimensional accuracies with commercially available Rapid Prototyping systems are usually measured in hundreds of microns. There is still some debate as to the necessity of low dimensional accuracies. It is clear, however, that surface finish and dimensional accuracy will play an important role in Rapid Tooling.

Hence it is a long term goal to produce a system to produce metal tools directly, without the use of costly equipment such as lasers, while improving standards of dimensional accuracy and surface finish. It is intended to do this using High Speed Selective Jet Electrodeposition.

2. ELECTRODEPOSITION

2.1 Conventional Electroplating

Electroplating is a process whereby two conducting electrodes are placed in a solution containing metal ions (electrolyte) and a potential applied between them. Metal ions are reduced at the surface of the negative electrode (cathode) and form a deposit of metal atoms on the surface. This grows at a rate proportional to the current passed. Metal ions from the positive electrode (anode) dissolve into the electrolyte.

In a simple plating system such as acid copper sulphate, if a current flows under an impressed potential between the electrodes, the current is carried through the electrolyte by all ions present in the electrolyte. Positively charged ions such as copper (Cu^{2+}) and hydroxonium ions (H_3O^+) move towards the cathode whilst sulphate ions (SO_4^{2-}) move towards the anode under the influence of the applied electric field. At the cathode surface under normal deposition potential, the copper ions alone are reduced to copper atoms by the excess electrons existing at the cathode as a result of the applied potential. The reduced copper atoms become incorporated onto the lattice structure of the electrode creating the plated layer. However, this process causes depletion of copper ions in the vicinity of the cathode as the rate of deposition tends to be significantly faster than their rate of replacement by migration. As a consequence, a concentration gradient develops which results in a diffusion flux across the gradient. This electrolyte layer adjacent to the electrode is referred to as the Diffusion Layer.

Fick's First Law of Diffusion states that the mass transport of ions by diffusion is proportional to the concentration gradient. However, if a sufficiently high current density exists at the electrode, the rate of deposition may be such that diffusion and migration cannot re-supply the metal ions fast enough to maintain an excess concentration of ions. As soon as metal ions reach the surface they are reduced and the metal ion concentration at the surface approaches zero. An illustration of the concentration profile of metal ions in an electrolyte adjacent to the cathode under these conditions is shown in Figure 1. The thickness of the diffusion layer (δ shown

Figure 1) is governed by both the hydrodynamic conditions of the electrolyte and the bulk metal ion concentration. Higher rates of bulk electrolyte movement result in a thinner diffusion layer. The maximum concentration gradient (c_b/δ) is thus governed by the hydrodynamic conditions and the bulk concentration and this determines the limiting current density i_L which can be expressed by equation [1].

$$i_L = \frac{- n F D c_b}{\delta (1 - t)}$$

[1]

n is the number of electronic charges carried by each metal ion, F is Faraday's Constant, D is the diffusion coefficient and the term $(1 - t)$ accounts for the migration of the ions under the electric field, where t is the transport number of the metal ion concerned.

In general, within electrolytes, metal ions are too small and highly charged to be energetically stable on their own. Hence they are usually surrounded by charged or polarised molecules. In an aqueous solution these are usually water molecules. Other species such as CN^- ions can be made to form complex molecules around the metal ions, often resulting in a net negative charge (in this case the complexes are still reduced at the cathode despite their electrostatic repulsion to it).

Usually the complexes adsorb onto the cathode (via weak Van der Waals type bonds). This arrangement allows for surface diffusion where the complexes move to more energetically favourable sites such as steps or holes. Finally, electrons are donated by the cathode and the metal ions are reduced to atoms which are then held by metallic bonds to the cathode.

The effect of surface diffusion aids the levelling of the deposit, with depressions being more energetically favourable sites than elevations.

2.2 High Speed Selective Jet Electrodeposition

High Speed Selective Jet Electrodeposition (HSSJE) is not a new technique. Patented by NASA in 1974, it uses a free standing jet of electrolyte impinging onto the cathode substrate as shown in Figure 2.

A current is passed from an anode which is placed upstream from the nozzle. Deposition occurs only in the impingement region and the immediately surrounding region as a result of the formation of an extremely thin radial wall jet layer of electrolyte. The electrical resistance of this wall jet is high in comparison to the impingement region so no deposition can occur there. As a result, jet deposition provides a means of maskless selective plating. Furthermore, due to the continual supply of fresh electrolyte and the hydrodynamic conditions created the thickness of the diffusion layer δ is smaller and hence the mass transport of metal ions to the surface of the cathode can be made substantially higher than in the case of traditional electroplating. Consequently, higher current densities and thus higher deposition rates can be achieved than is conventionally possible. A more detailed discussion of the process is given by Bocking(1) and Chen(2).

However, at the very high current densities found with HSSJE, the time allowed for surface diffusion is much lower and deposits can be rough, porous, hard and contain higher internal stresses.

Also, since the diffusion layer is much thinner any small elevation in the cathode surface will protrude into a region of higher metal ion concentration. This produces a locally enhanced current density and the protrusion is magnified. By this effect, the maximum current density that produces a good deposit is set well below the maximum current density as described earlier.

By moving the nozzle in relation to the substrate whilst deposition is occurring, it is possible to selectively "write" tracks and patterns at relatively fast rates without the need for masking the substrate.

2.3 Properties of Electrolytes

In order to facilitate higher speeds of deposition, higher concentrations of metal ions are often used. In addition, higher temperatures increase the diffusion rate of metal ions and jet deposition benefits from electrolytes operating at elevated temperatures. However, many conventional plating electrolytes are formulated using constituents that are corrosive to the metal that is deposited. Consequently, deposits produced by HSSJE may be re-dissolved by the flowing electrolyte during "writing" operations. Therefore, electrolytes often have to be especially formulated to minimise or eliminate this effect.

As organic additives used in conventional techniques (brighteners, levellers etc.) are usually adsorbed onto the cathode (a time dependant phenomenon), their fraction of incorporation is largely dependent on the current density. High current densities and thus high rates of deposition will result in less incorporation into the deposit. Hence at high current densities these additives are less effective.

Conventional electrolytes are very different from good HSSJE electrolytes, to the extent that HSSJE electrolytes often will not work in conventional plating baths.

3. POTENTIAL APPLICATIONS WITHIN RAPID PROTOTYPING AND TOOLING

The potential applications for this technology include :

1) To directly produce metal tools in nickel or copper which can be used for pressing or plastic injection moulding or produce an EDM anode for harder tooling.

2) To produce the protruding features of a tool that would be difficult to machine.

3) To produce specialised printed circuit boards, e.g. microwave circuits(3).

3.1 Difficulties

1) Due to the low currents (despite the high current densities) associated with HSSJE build rates are measured in grams per hour. Hence it would take an extremely long time to build a large metal tool. Either improvements must be made in the deposition rates or multiple nozzles must be used, possibly with varying apertures for combined dimensional accuracy and speed.

2) Due to the factors that amplify surface imperfections, feedback systems must be developed to build large deposits. However, provided the parts are reasonably shallow, this could be omitted.

3) Support structures may prove difficult to produce. Overhangs could be produced from a nozzle to substrate angle of less than 90°. The mass transfer characteristics of an oblique jet have

been previously documented by Chin & Agarwal(4).This method may not be amenable to the usual CAD $\Rightarrow$ STL $\Rightarrow$ SLI $\Rightarrow$ vector file process found with other Rapid Prototyping systems.

4) Some of the chemicals used can be harmful. In particular, if these chemicals are forced through a small nozzle onto a substrate at high velocities, hazardous spray can be evolved.

4. EXAMPLE APPLICATION - PRESS TOOLING

It was decided to attempt to produce the male part of a press tool. The hardness has to be high and hence nickel was selected for the deposit. It was decided to produce a protruding circle, with height of about 0.5 mm. Little work has been carried out previously on HSSJE of nickel, although the deposition of gold and gold alloys has been ostensibly studied. Therefore, this application made use of a conventional nickel electrolyte used for electroforming. It should be noted that this electrolyte may not be the optimum for use with HSSJE (see 2.3 above).

4.1 Experimental Conditions

The conditions used are given here (unless otherwise specified):

Electrolyte:	600 gl^{-1} nickel sulphamate
	30 gl^{-1} boric acid
	@ 52 - 55 °C
Nozzle orifice diameter:	ϕ400 μm
Nozzle to substrate distance:	0.5 $\pm$ 0.1 mm
Electrolyte velocity:	15 - 20 ms^{-1}

The nozzle could be moved parallel and perpendicular to the substrate using stepper motors controlled from a PC. The apparatus was configured with the substrate mounted vertically and the nozzle horizontally.

4.1 Finding the correct current density

Since the deposition rate is governed by the current passed, a higher current will produce a lower build time. However, high current densities can produce rough deposits. Hence it was necessary to determine an optimum current for future use.

Three pieces of glass, sputter coated with ~1000Å of gold (giving an almost even <111> orientation) were jet plated with 10, 14 and 14 separate spots of differing current densities but all with the same charge (0.078 C) passed. Each sample had spots with current densities of 0.25, 0.75, 1, 2, 3, 4, 5, 6, 7, 8 Acm^{-2} but with extra spots at 9, 10, 11 and 12 Acm^{-2} for the second and third samples. The samples were then washed in deionised water and dried in an oil free air jet. By examination of the quality of the deposit under an optical microscope, it was seen that at very low current densities (0.75 Acm^{-2}), the deposits were highly stressed, with the edges pulling the very thin gold sputtered layer off the glass (see Figure 3). At high current densities the deposits become rough and nodular (see Figure 4). It was decided that 7 - 8 Acm^{-2} was the maximum useful current density (see figures 5,6,7 & 8).

4.2 Finding the correct sweep speed

When producing deposited tracks using HSSJE, the tracking speed of the nozzle has an upper limit. This is due to the finite time it takes for both the reduction process and deposit nucleation to occur. In order for electron transfer to occur, the metal ions need to find energetically suitable sites on the surface and, therefore, some surface diffusion of the ions has to take place. This takes a short but finite time. Nucleation is the initial stages of deposit formation in which the reduced metal atoms diffuse together to form "islands" of metal that grow both vertically and laterally. This also requires a small but finite time. If the nozzle movement is too rapid then there is insufficient time for these processes to occur adequately and a significant reduction in actual deposition rate is observed. Hence it was necessary to determine the range of nozzle sweep speeds that can be used.

A piece of polished mild steel was immersed in an ultrasonic cleaner, then in a cathodic alkali cleaner, then in a cathodic cyanide cleaner, then a very light etch in 20% HCl and then plated in nickel sulphamate for ~ 10 minutes. 12 separate lines of equal length were jet plated onto this substrate using one pass at varying nozzle sweep speeds and varying current densities. At 7 Acm^{-2}; 0.1, 0.2, 0.4, 0.6, 0.8 and 1 mms^{-1} and at 8 Acm^{-2} 0.1, 0.2, 0.4 and 0.6 mms^{-1}. Using very rough measurements of the profiles of deposits produced at 8 Acm^{-2}, a 12 μm high deposit was produced at 0.1 mms^{-1} and a 2 μm high deposit was produced at 0.6 mms^{-1}. This indicated that it was likely that there was no difference in cathodic efficiency or sharpening of the deposit with increasing nozzle sweep speeds (as the deposit height was inversely proportional to the sweep speed). Hence with the sweep speeds studied, no upper limit was found. Optical microscopic examination of the deposit quality confirmed this as only one small crack was observed in all the deposits (see Figure 9). This crack had an unknown cause, possibly residual stress.

4. 3 Creating Nickel Deposits

On a piece of gauge plate with a ground surface, chemically treated as the piece of mild steel in experiment 2 (with a nickel surface), a circle of radius 58 mm was produced with a sweep speed of 4 mms^{-1} and a current density of 7 Acm^{-2}. This appeared to have cracks which were oriented with respect to the finish of the gauge plate. This indicated that the plating was amplifying any surface imperfections. This can be seen in Figure 10 where surface scratches on a polished substrate are continued in the deposit. Hence a ground surface would be unsatisfactory. Also, as one full circle was completed and more nickel was deposited on top of previously jet deposited nickel, it was noticed that the adhesion of the second layer to the first was poor. This could be seen when a piece of adhesive tape applied to the deposits was pulled away taking with it the upper deposit but leaving the lower one. This was attributed to the nickel passivating, the resultant oxidised layer not being conducive to good adhesion (see Figure 11). An experiment was performed under similar conditions to the previous one except at a higher sweep speed of 6 mms^{-1}, so as to complete a full circle before the nickel had sufficient time to fully passivate. However, the nickel still did not adhere to lower layers of jet plated nickel. It was also noted that deposits exhibited high internal tensile stresses, often with the thin outer edges of the deposits curled up away from the substrate (see Figure 12).

Attempts were made to solve these problems. A hard cobalt-gold flash was put down over the nickel plate to prevent tarnishing of the nickel plate. This gave good adhesion of the first pass of jet plated nickel with no lifting of the edges (see Figure 13). Hence the adhesion was increased enough to stop the internal stress from lifting the edges off the substrate. However, a second pass on top of the first pass still gave poor adhesion.

An attempt to deposit a higher deposit in one pass was made but unfortunately the apparatus was limited to a minimum sweep speed of 0.05 mms^{-1}, limiting the final height of the deposit.

4.4 Creating deposits for profile measurement

As gold and gold alloy jet deposits have been extensively studied, it was decided to produce test samples of a gold/nickel alloy. Deposits of this type have been shown not to suffer from surface passivation when overwriting as was found in the case of pure nickel.

Two samples were produced with circles of gold deposits ϕ58 mm and with measurable heights to give representative profiles, with a citrate-gold nickel alloy electrolyte producing a deposit with a Vickers microhardness of 190 and a nickel content of 0.3%. The conditions were:

Reynolds number:		$10,000 \pm 500$
Nozzle orifice diameter:		ϕ 0.36 mm
Nozzle - substrate distance:		1.0 ± 0.2 mm
Current density:	Sample 1:	5 Acm^{-2}
	Sample 2:	7 Acm^{-2}
Number of segments in circle	Sample 1:	360
	Sample 2:	3600
Sweep speed		0.75 mms^{-1}
Time		7 hours.

The first sample showed a good quality deposit as seen in Figure 14. This deposit is 0.10 mm high. However, due to the low resolution of the controlling software (360 segments in the circle), the nozzle dwelled for a finite time at the ends of the segments, forming lumps in the deposit. These can just be made out in Figure 15. Using a higher resolution of controlling software, a better quality of deposit was produced as can be seen in Figure 16. This deposit is 0.17 mm high. A profile of this deposit was measured and is shown in Figure 17 along with a computer generated profile of form $y = A e^{(-Bx^2)}$ with a value of B of 27.7 mm^{-2}.

However, a sharper deposit will be obtained when the deposit grows by a significant fraction of the nozzle to substrate distance as the top of the deposit will attract an increased share of the current. Hence, movement of the nozzle away from the substrate during the build may make the deposit less sharp and alter the shape of the deposit.

The deposit seems to lean slightly. This may be due to the nozzle leaning but is more likely to be due to the substrate being bent when a profile was measured or the profile not being aligned properly when digitised.

5. CONCLUSIONS

This work demonstrates that by using HSSJE, it is possible to produce fine definition press tooling. Whilst gold alloy deposits demonstrate the feasibility, problems remain with the conventional nickel electrolyte used in this study and the deposits produced from it.

In order to produce press tooling using HSSJE nickel work has to be done on reducing the residual stress of the deposit, to prevent passivation to allow multiple passes or reduce the minimum lateral speed of the nozzle and to develop procedures to produce profiles to order.

The first problem may be solved by altering the composition of the electrolyte. More work must be carried out on this problem.

The difficulty of passivation may again be solved by altering the electrolyte or by jet depositing copper immediately over the nickel deposit but this presents problems in keeping the two electrolytes separate. Again, more work is necessary.

The problem of producing a tailor-made profile can either be solved by very complex mathematical modelling or by empirical evidence and experience, the former involving many variables and the latter being limited in application.

6. FIGURES

concentration

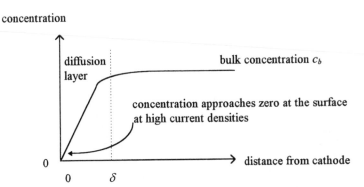

Figure 1. The concentration profile of metal ions in the diffusion layer.

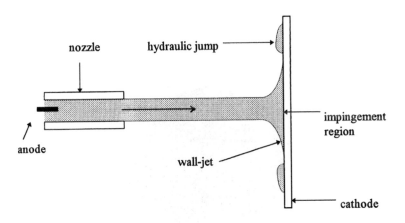

Figure 2. High Speed Selective Jet Electrodeposition using an unsubmerged jet.

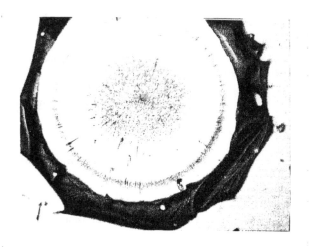

Figure 3. A nickel spot using 0.078 C of charge produced at 1.0 Acm⁻². The deposit has a fine surface although the edges have peeled away from the substrate. x 107.

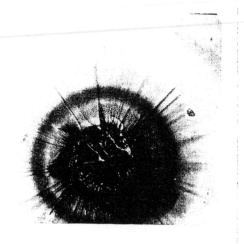

Figure 4. A nickel spot using 0.078 C of charge produced at 9 Acm⁻². The surface of the deposit is very coarse. x 107.

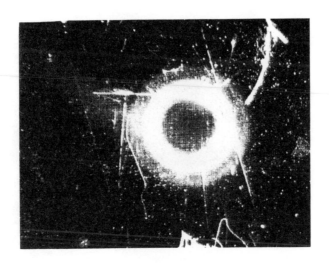

Figure 5. A nickel spot using 0.078 C of charge produced at 7 Acm^{-2}. No coarse surface morphology can be seen. x 67.

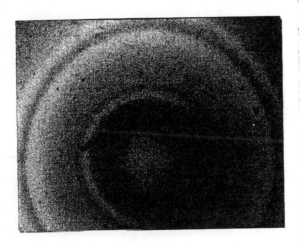

Figure 6. A nickel spot using 0.078 C of charge produced at 4 Acm^{-2}. The surface is very fine. x 107.

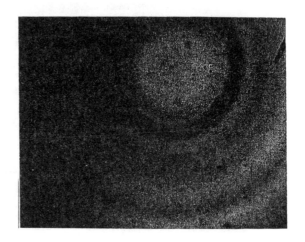

Figure 7. A nickel spot using 0.078 C of charge produced at 7 Acm^{-2}. The surface is very fine. x 107.

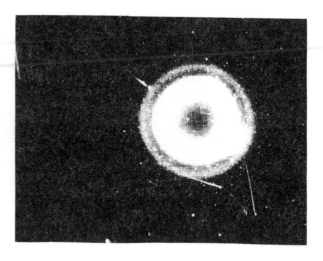

Figure 8. A nickel spot using 0.078 C of charge produced at 6 Acm^{-2}. x 67.

Figure 9. A nickel ridge produced at 8 Acm^{-2} and a sweep speed of 0.1 mms^{-1}. x 67.

Figure 10. A nickel ridge. The surface finish of the polished substrate remains on the surface of the deposit. x 67.

Figure 11. A nickel ridge with one deposit on top of a second one. x 67.

Figure 12. A nickel ridge. The thin edges of the deposit have curled away from the substrate. x 67.

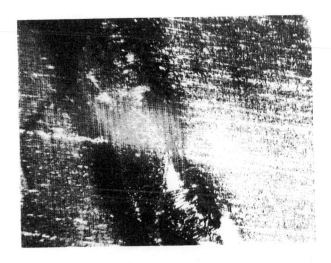

Figure 13. A nickel ridge. An inert coating of gold prevents curling at the edges. x 67.

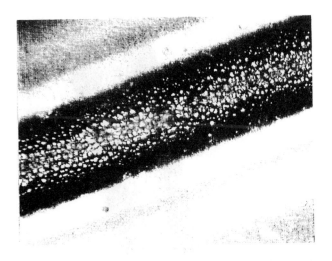

Figure 14. A gold/nickel alloy ridge produced at 5 Acm^{-2} and 0.75 mms^{-1}. x 67.

Figure 15. A gold/nickel alloy ridge produced at 5 Acm^{-2} and 0.75 mms^{-1}. Two 'lumps' are visible which mark the ends of the segments used by the controlling software to produce a circle. x 67.

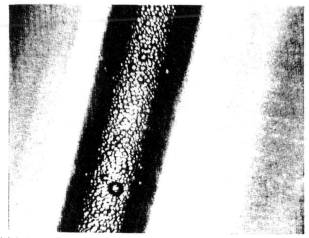

Figure 16. A gold/nickel alloy ridge produced at 7 Acm^{-2} and 0.75 mms^{-1}. The nodule seen at one end is the largest one seen in the whole circle. x 67.

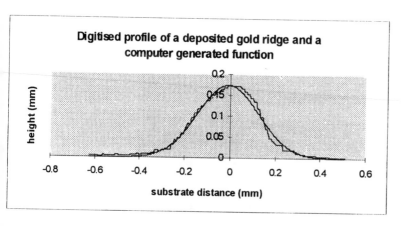

Figure 17. The profile of a jetted gold/nickel alloy deposit. The smooth line is a computer generated function of the form $y = A e^{(-Bx^2)}$.

7. REFERENCES

(1) BOCKING, C.E., <u>High Speed Selective Jet Electrodeposition of Gold and Gold Alloys using Single Circular Jets</u>, Doctoral Thesis, Institute of Polymer and Materials Engineering, Loughborough University of Technology, 1994.

(2) CHEN, T-J., <u>Selective Jet Plating</u>, Doctoral Thesis, University of Illinois, 1981.

(3) BOCKING, C.E., High Speed Selective Jet Electrodeposition, <u>Trans. Inst. Metal Finish.</u>, April 1991, **69**(4), pp.119-127.

(4) CHIN, D-T., AGARWAL, M., Mass Transfer from an Oblique Impinging Slot Jet, <u>J. Electrochem. Soc.</u>, September 1991, **138** No. 9, pp.2643-2650.

An investigation into the formation of bubbles in stereolithography parts made from Ciba Geigy SL5170 epoxy photopolymer on the SLA190

D HALE and **G BENNETT**
Buckinghamshire College of Higher Education, UK

SYNOPSIS

The presence of bubbles within the solidified structure of stereolithography parts, particularly parts made on the SLA190 stereolithography machine using Ciba Geigy SL5170 photopolymer resin is common. Such bubbles typically appear in 'columns' and their presence is frequently detrimental to the surface finish and strength of the parts.

An investigation is reported to determine the causes of bubble formation, and their frequency of occurrence. This investigation studies the formation of bubbles as a function of part geometry, wall aspect ratios and build parameters.

Different strategies of bubble formation prevention have been investigated. Their relative merits and successes are reported.

1 INTRODUCTION

The Centre for Rapid Design and Manufacture at the Buckinghamshire College has been using a 3D Systems SLA190 stereolithography machine with Ciba Geigy SL5170 epoxy resin since April 1995. The majority of the parts are built using the ACES build style. One of the most serious problems that has been encountered during its operation has been the occurrence of bubbles in the parts.

The SLA190 is based on the SLA250, but has a smaller resin tank, and significantly, no recoating arm. In practice, this means that thin wall sections are quicker to build (as no time is required for a wiper blade to operate). However, thick walled sections take significantly longer to build, due to the extended wait time necessary for the liquid resin to level without recoating arm assistance. The arm also has a dramatic effect on the presence of bubbles in finished parts, which will be discussed later.

The significance of the presence of the presence of bubbles in parts has recently been surveyed(1). 51% of those participating in the survey listed the poor physical properties of the materials used as a serious problem with the system, and 26% listed the poor surface finish as a serious problem. Both of these properties are directly influenced by bubbles in the parts.

2 RESIN PROPERTIES

The resin used for these investigations was the epoxy based Ciba Geigy SL5170.

In comparison with the Ciba Geigy SL5177 acrylate based resin (2,3), the epoxy has a tensile strength 70% higher than the acrylate, an elastic modulus 120% higher, and a similar impact strength and hardness. The shrinkage of the epoxy resin during polymerisation is lower than that of the acrylate, which results in less internal stresses in the part, and therefore less warp and curl.

The epoxy has a viscosity of about 10% of that for the acrylates, which gives quicker recoating times. However, the epoxy has a higher critical exposure (E_c) and lower depth of penetration (D_p) than the acrlylates, which means that more laser energy is required to make parts in epoxy.

The epoxy resins also allow the use of the ACES build style, which has a higher cross hatch density, giving a more thorough cure than is possible using the STARWEAVE build style used with the acrylates. Hence less time is needed in the post curing apparatus.

3 BUILD PROBLEMS

The appearance of bubbles is due to a combination of: the 'Deep Dip' recoating system used on 3D Systems machines; the viscosity of the resin; and the lack of a wiper blade.

The formation of bubbles in parts made by stereolithography is a problem for several reasons.

i)   The strength of the part is decreased. Furthermore the bubbles can join together to form large columns in the part, which significantly weakens the part. Individual bubbles will act as stress concentrators.

ii)   Bubbles can break through the surface of thin walled parts. This can create problems if the part is to be used in further processes such as investment casting or silicon rubber moulding, or if the part is to be used directly as a mould.

iii)   Ridges produced by 'lines' of bubbles can change the dimensional accuracy of the part.

vi)   The surface finish of thin walls is poor as the bubbles tend to bow parts outwards.

4 EXPERIMENTAL TESTS

To investigate these problems, a set of test pieces were made, as shown in figure 1. The same build was carried out several times to see how repeatable the results were.

The parts were built using standard build conditions of z velocity = 0.2, z acceleration = 0.2, post dip delay = 1 seconds, pre dip delay = 6 seconds and z wait = 60 seconds.

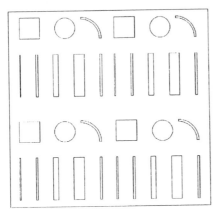 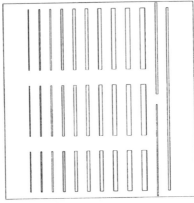

Front of tank   Front of tank

Figure 1. Plan view of build table   Figure 2. Plan view of build table showing
(190x190mm) showing the position of   the position of 'test-2' parts.
'test-1' parts.

It was suggested that the bubbles may be being formed when the platform was initially lowered into the resin before each build, and that this problem could be avoided by wiping the platform several times before the build was started. This was done before a third set of the test pieces were made.

From these initial results, a second set of test pieces were built (see figure 2), with parts of different lengths and widths (x,y plane), to see if there was a specific length to width ratio (aspect ratio) that reduces the occurrence of bubbles.

A second set of these parts were also made on an SLA250, using the wiper blade, with process conditions as close as possible to those used on the SLA190.

The length of the bubble column was measured on the test pieces to give an indication of the severity of the problem.

The diameter of the 'test-2' parts bubbles measured on a travelling microscope to determine if there was a relationship between the wall thickness of the part and the size of the bubble.

A third set of test pieces were designed, as shown in figure 3, all with the same cross section (x,y plane), and differing aspect ratios to try to isolate the two parameters.

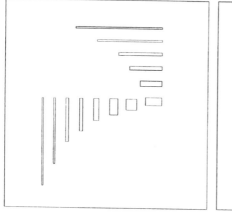

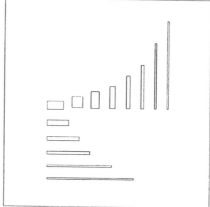

Front of tank   Front of tank

Figure 3. Plan of build table showing the   Figure 4. Plan of build table showing the
position of 'test-3' parts   position of 'test-3C' parts. (see section 7)

The 'test-3' parts were made again. The bottom half of the parts were built with a post dip delay of 99 seconds, and the top half of the parts were built with a post dip delay of 50 seconds. This was done to see if the occurrence of bubbles would be affected by allowing the bubbles more time to burst during this stage.

A video recording was taken of several different parts being made, in order to observe the recoating process more closely.

Tensile tests were carried out on some of the parts which contained bubbles in them to investigate the strength of the parts. The tests were carried out on a Lloyd instruments T30K tensile testing machine.

Photographs were taken of some of the parts with bubbles on a stereo dynoscope to examine the form that the bubbles take when the part is solidified.

5 RESULTS

The results from the first set of test results (see figure 5), showed that the bubbles were formed predominantly in the thin wall sections. The bubbles always formed in the centre of each layer, hence bubble columns captured in parts are not neccessarily vertical, where the wave fronts met during the recoating process. The thin walled arcs and angled pieces also formed bubbles in the same way. The square and cylindrical parts did not have bubbles present.

The position of the sets of parts on the platform did not have an effect on the occurrence of bubbles, other than the fact that the last set of pieces had a longer wait time, and so the bubbles were more likely to have burst.

Wiping the build platform after it was initially dipped into the resin had no measurable effect on the occurrence of bubbles in the part.

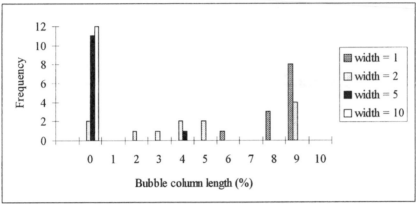

Figure 5. Plot of frequency of bubbles against size of bubbles for 'test-1' parts.

The relationship between the part aspect ratio and the occurrence of bubbles can be looked at more closely in the results from the 'test-2' parts. These show that parts with a high aspect ratio are more likely to have more bubbles, as shown in figure 6.

The second pattern of test parts was made on an SLA250, with a wiper blade, and did not contain any significant number of bubbles. The bubbles were either burst as the blade moved over the parts during recoating, or were pushed away from the part, before the laser wrote onto the surface.

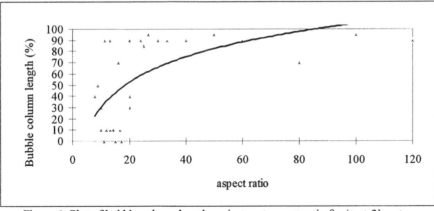

Figure 6. Plot of bubble column length against part aspect ratio for 'test-2' parts.

The wall thickness does not have any great effect on the size of the bubbles trapped in the solid parts, as shown below.

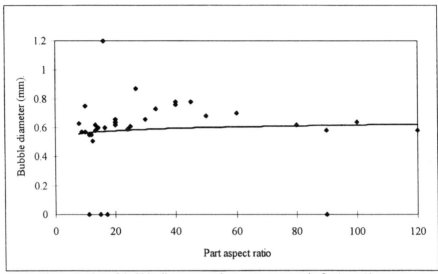

Figure 7. Plot of bubble diameter against part aspect ratio for 'test-2' parts.

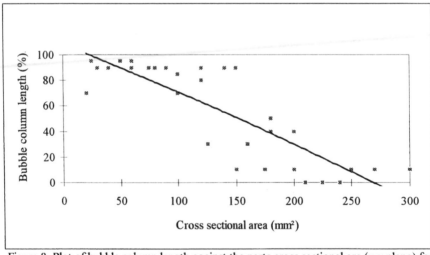

Figure 8. Plot of bubble column length against the parts cross sectional are (x,y plane) for 'test-2' parts.

The 'test-3' parts have a constant surface area and different aspect ratios. In these parts, the aspect ratio of the part did not have any affect on the occurrence of bubbles.

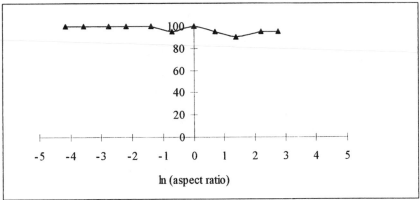

Figure 9. Plot of bubble column length against part aspect ratio for 'test-3' parts.

The second set of 'test-3' parts, which had a height of 10mm, were made with a post dip delay of 99 seconds for the lower 5mm of the part, and a post dip delay of 50 seconds for the upper 5mm of the parts. More bubbles occurred in the parts with an aspect ratio close to one in the second half, as they have less time to burst naturally during the wait time.

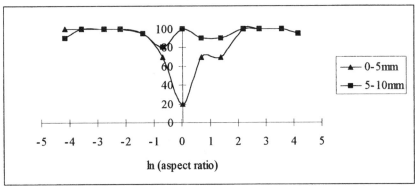

Figure 10. Plot of bubble column length against part aspect ratio for parts with a post dip delay for the first 5 mm of the part = 99 seconds, and equal to 50 seconds for the second 5 mm of the part.

Analysis of the video recording of the recoating process showed that the bubbles always formed as the platform was lowered, and the part recoated. As the part is recoated, the resin does not flow evenly over the part, but because of its high viscosity, the resin moves in to cover the part from two sides. The waves move in a direction parallel to the longest edges unless the part is a square or cylinder, where the resin moves in from all sides simultaneously. The bubbles form where the waves of recoating resin meet.

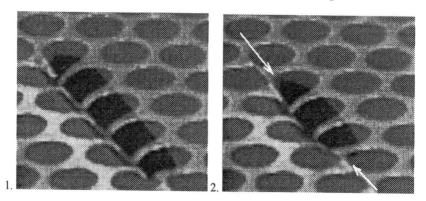

Figure 11. Video image of the recoating of rectangular and square parts, showing the resin waves moving together. In the background, the build platform can be seen.

As the resin moves over the part, its viscosity means that a curved wave front is formed (see figure 12). When the two waves meet, they trap air between them forming a bubble. The bubble remains trapped on the part, as immersion occurs. During descent, bubbles typically detach from the part and float to the surface. When the part resurfaces, the bubble is captured. If a captured bubble is illuminated by the laser, it typically bursts, leaving a hollow crater. After the first bubble has been solidified, then the hollow already formed promotes the formation of further bubbles, which in turn are solidified, and so join together to form a full bubble in the final part.

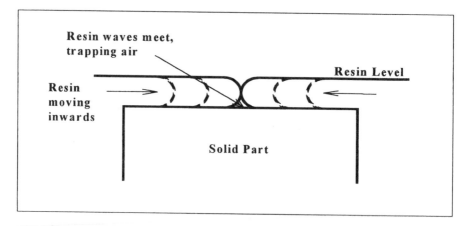

Figure 12. Diagram showing the resin trapping air during recoating.

The decrease in strength due to the bubbles was shown by carrying out tensile tests on some of the parts.

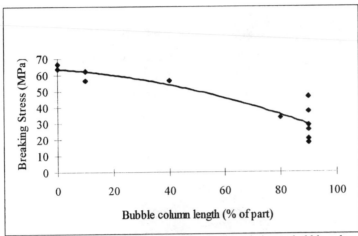

Figure 13. Plot showing tensile strength of parts with relation to bubble column length, shown as a percentage of the part.

Photographs taken on a stereo dynoscope of several of the parts show the bubbles in more detail.

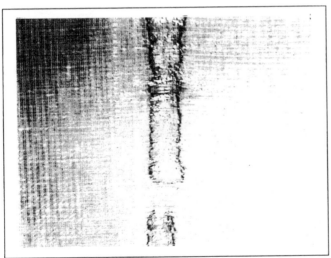

Figure 14. Photograph (x10) of part 28A from 'test-1'(a rectangular part, 40x1x10mm).

This shows a continuous column of bubbles which have merged together to form a hollow cylinder in the part.

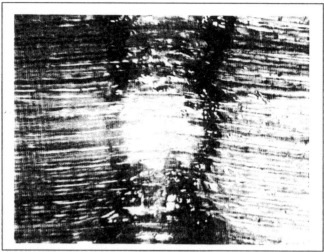

Figure 15. Photograph (x40) of part 28A from 'test-1' (a rectangular part, 40x1x10mm).

This shows the bowing up of the part near to the bubble, which causes dimensional inaccuracies. This is due to the resin around the bubble being raised when the laser writes over it.

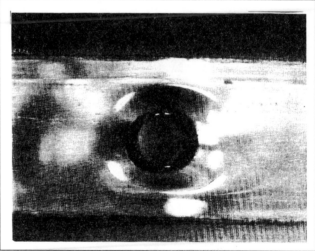

Figure 16. Photograph (x30) of part 29A from 'test-1' (a rectangular part, 40x2x10mm).

This photograph is a plan view, showing the top of the exposed hollow cylinder made by the bubbles. This would need to be sealed if the part were being used for investment casting or silicon rubber moulding.

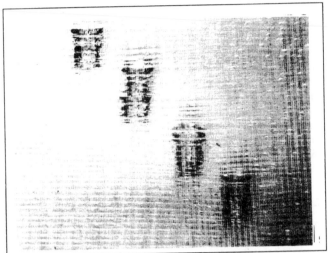

Figure 17. A photograph (x10) of part 32A in 'test-1' (angle piece 40x2x10mm).

This shows the bubbles moving across the part as the point where the resin waves meet moves to the left. It is also seen that the bubbles form cylinders with flat tops and bottoms, rather than a curved top to the void.

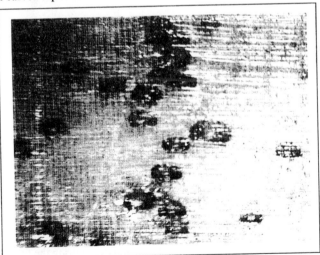

Figure 18. Photograph of part (x10) 13A in 'test-3' (a rectangle 80x1.25x10mm).

This photograph shows that the bubbles do not always form in columns. This may be due to the resin recoating at different speeds from each side of the part, or starting to recoat at different times, which will mean that the resin meets at different points, and so the bubbles form at different points.

6 DEVELOPMENT OF BLOWER

As the bubbles always form at the surface of the resin, it was thought that they could be pushed away from the parts by blowing air across the surface of the resin.

A plenum chamber and nozzle were designed to hook onto the front of the resin tank and hold a small motor. This box and nozzle were made on the SLA in several parts and assembled using screws and metal inserts.. When the plenum chamber and nozzle were assembled, this created a slot which fitted over the tank and positioned the bottom of the nozzle 2mm above the resin level. A 12V dc axial fan was attached to the base of the plenum chamber using screws and and metal inserts in the SLA part.

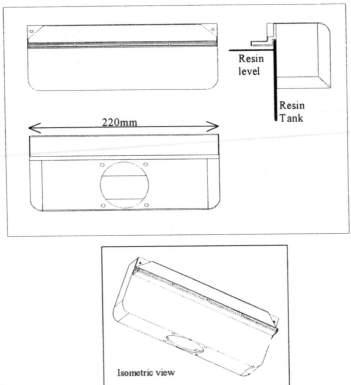

Figure 19. Line drawings of the assembled box and nozzle, showing the fitting point for the fan on the base.

7 RESULTS OF BLOWER OPERATION

The fan was initially tested on the 'test-3' parts, using a 60 second wait time and the blower was switched on from 4 seconds into the pre dip to 40 seconds into the z wait time. This showed that the majority of the bubbles were moved during the post dip and the start of the z wait time.

Taking this into account, the parts were built again ('test-3B') using a post dip delay of 15 seconds and a z wait of 15 seconds. The blower was left on from 4 seconds into the pre dip delay to 2 seconds into the z wait. The parts closest to the blower were perpendicular to the air flow, and did not have any bubbles in them. However, the parts that were furthest from the blower still had bubbles present in them. This was either due to the air not disturbing the bubbles sufficiently because air stream did not reach the back of the tank, or because the parts were parallel to the front of the tank, the parts were being sheltered from the airflow by other parts.

Therefore, the 'test-3' parts were rearranged so that the parts that were perpendicular to the airflow were near the front of the tank, and the parts that were parallel to the airflow, were at the back of the tank. This was labelled 'test-3C', as seen in figure 4.

Both the sets of parts that were made with the blower had a significantly lower density of bubbles in them, as can be seen in figure 20. The set of parts that were parallel with the airflow had less bubbles than the other set, which shows that the distance from the blower is significant, as well as the orientation and spacing of the parts in the tank.

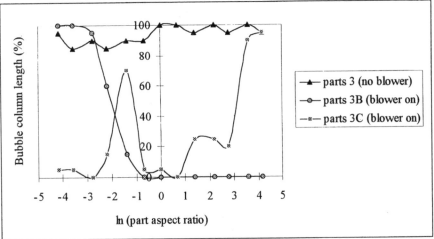

Figure 20. Plot of bubble column length as a percentage of the part height and part aspect ratio for different parts made with and without the blower.

The relation between the distance of the part from the front of the platform, where the blower is, was investigated by making a set of parts, that were all the same size, positioned at different distances from the front of the tank, with some of the parts perpendicular to the blower and some parallel.

The relationship between the distance of the part from the blower and the number of bubbles in the part is clearly shown in figure 21. There is an approximately linear relationship between the distance of the part away from the front of the blower and the number of bubbles in the part. This is due to the velocity of the air being higher at the front of the tank, so having sufficient force to blow the bubbles away from the surface of the parts. The relationship seen is approximately the same for parts that are built parallel or perpendicular to the front of the tank, if the parts at the back are not sheltered from the effect of the blower.

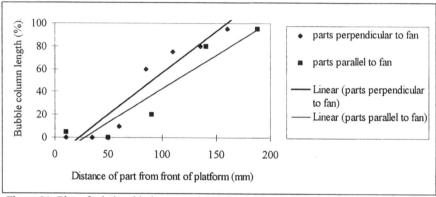

Figure 21. Plot of relationship between the distance of the centre of the parts to the front of the platform, and the number of bubbles in the parts.

The parts that were made with the blower were measured and compared to the identical parts made without the blower to check the dimensional accuracy. The blower had no measurable effect on the accuracy of the parts.

8 CONCLUSIONS

Bubbles have been seen to be a significant problem on parts made on the 3D Systems SLA190 with the Ciba Geigy SL5170 epoxy based resin. The problem is eliminated on larger machines such as the SLA250 and SLA500 that use a wiper arm during the recoating phase of the process.

The blower has been shown to be an effective solution to the problem of bubbles. As the blower is turned off during most of the z wait time of the process, it does not have any measurable effect on the accuracy of the parts.

REFERENCES

1   'Rapid Prototyping Report' Vol. 5, No. 3, March 1995.
2   'Rapid Prototyping Report' Vol. 4, No. 10, October 1994.
3   '3D Systems Cibatool Resin Handbook'

ptimising arc–sprayed metal tooling for injection oulding

EGAL and R C COBB
versity of Nottingham, UK

1. ABSTRACT

Arc-sprayed metal tooling has been a commercial process since the early 1970's and it enables the rapid fabrication of moulds for a variety of different manufacturing processes including injection moulding, reaction injection moulding (RIM), reaction reinforced injection moulding (RRIM), and vacuum forming. However, the process suffers from limitations such as short tool life and high wear rates and also the limited product complexity that can be accommodated by such a process.

Rapid prototyping technologies (RPTs) have been in existence since the late 1980's and these processes provide physical models for evaluation in terms of aesthetics, dimensions and material volume. However, for many manufacturers this is not sufficient and so a rapid method for producing prototypes using the production process and in the production material is required. The advent of rapid prototyping technologies has provided a new impetus for the sprayed metal tooling process due to the ability to manufacture patterns rapidly.

Within the framework of a Brite Euram II CRAFT project entitled 'Low Cost, Short Run and Prototype Tooling' (Project No. CR-1357-91), the process of metal arc-spraying is under investigation as a 'rapid' injection mould tooling process. The primary objective is to optimise the metal arc-spraying process and to develop new raw materials so that more durable and complex injection mould tools may be produced. This paper describes the process of sprayed metal tool-making and identifies its advantages and disadvantages in comparison with conventional tool-making processes and materials. The moulding process of a two-impression sprayed metal injection mould is also outlined.

2. INTRODUCTION

Arc-sprayed metal tooling was developed in the early 1970's (1). Prior to this most sprayed metal tooling had been produced using either a low melting point spray gun or a flame spray gun. Arc spraying has several advantages over these two processes, it can be used to spray harder metals than the low melting point spray gun, such as bronzes and steels and also it has a higher deposition rate and thermal efficiency than flame spraying. The main benefit of arc-sprayed metal tooling is that, especially when compared with conventional tool-making processes, it is relatively simple and inexpensive.

A sprayed metal tool can be thought of as a three part 'system' (1) :-

- BOLSTER - The bolster, or cavity plate, is used to contain the mould cavity. It has to withstand the clamping force of the injection moulding machine and maintain the alignment of both mould halves. Aluminium is the most common metal used for the bolster.

- SPRAYED METAL SHELL - The shell is a negative replication of the form of the pattern. The shell forms the face of the tool and hence must have a good surface finish. The shell is usually sprayed to a thickness of 1.5 to 2 mm using fine atomising pressures to replicate the pattern.

- BACKING SYSTEM - A strong backing system is required to support the sprayed metal shell inside the bolster. The backing must withstand the injection pressure and temperature and should also be compatible with the shell, that is having a similar coefficient of thermal expansion. There should also be good bonding between the backing material and the metal shell. Typically, the backing system is of a metal filled resin.

The fundamental requirement of any sprayed tool is an accurate pattern. This has previously been fabricated from materials such as wood, plaster, resin and occasionally metal. Pattern production has traditionally occupied a significant proportion of the total tool manufacturing time and cost, is relatively inaccurate and relies upon simple patterns. To make sprayed metal tooling more competitive with alternative tool-making systems such as machined aluminium, resin tooling etc., a faster method for generating patterns is desired. Rapid Prototyping (RP) has developed over the past decade into a reliable method of rapidly producing highly accurate, complex and low cost prototype models.

The technology of rapid prototyping is very quickly finding beneficial application areas within manufacturing organisations world-wide and presently a great deal of interest is being shown in further applications of rapid prototyping techniques (2,3). Increasingly, manufacturers are looking towards using prototypes to produce tooling.

The objective of this paper is to discuss the advantages and disadvantages of sprayed metal tool-making in comparison with conventional tool-making processes and materials. To support this, the moulding process of a two-impression sprayed metal injection mould is described.

3. RAPID TOOLING

With the development of RP the concept of Rapid Tooling has arrived. Here, both prototype and production tooling may be produced by using rapid prototyping techniques to manufacture the pattern, combined with low cost tooling methods. It may eventually prove commercially viable to produce some tooling directly (4), that is producing the tooling directly using a particular rapid prototyping technique, but most tooling is produced indirectly, for example using a stereolithography component as a master pattern from which a silicone rubber mould can be produced which may then be used for the vacuum casting process to produce prototype parts.

Figures 1 (a) and (b) summarise the main commercial prototype tooling processes and give an indication of why so much interest is being shown in rapid prototyping and tooling. The numbers in brackets are typical quantities achievable for the particular tooling system.

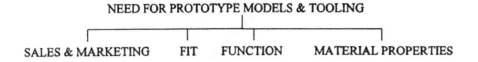

NEED FOR PROTOTYPE MODELS & TOOLING

SALES & MARKETING   FIT   FUNCTION   MATERIAL PROPERTIES

Fig 1 (a) - Applications of rapid prototype models

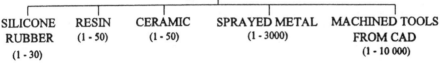

ADVENT OF COMPUTER-GENERATED MODELS (FAST & LESS EXPENSIVE)

ECONOMICAL USE OF PATTERNS TO FABRICATE TOOLING
(PREVIOUSLY TOO COSTLY)

MAIN PROTOTYPE TOOLING PROCESSES

| SILICONE RUBBER (1 - 30) | RESIN (1 - 50) | CERAMIC (1 - 50) | SPRAYED METAL (1 - 3000) | MACHINED TOOLS FROM CAD (1 - 10 000) |

Fig 1 (b) - Summary of rapid tooling processes

The UK tooling market has been valued at approximately £386 million, with lead times of anything from 1 to 26 weeks (5). Rapid tooling could reduce costs by at least 20% and tooling lead times by 20-80%. Typically, sprayed metal tooling offers significant time and cost savings of at least 50 % over conventional machined tooling (6,7) and usually the cost saving achieved is greater than the time saving.

4   SPRAYED METAL TOOLING

4.1   Advantages

Most of the research and development with sprayed metal tooling has concentrated on mould making for applications such as vacuum forming and injection moulding. The electric arc spraying process (see Figure 2) is by far the most suitable for the production of metal sprayed tooling mainly because it avoids excessive heat build up in the pattern and mould distortion is minimised (8). It also has the lowest relative cost of all the thermal spraying processes, a relatively high spraying rate (9) and is the most widely used metal spraying process for producing moulds. The number of parts obtainable from a sprayed metal tool is generally quoted as between 1 and 3000 (1) but this can be severely limited by the part geometry and material type especially when using high pressure and high temperature polymers.

191

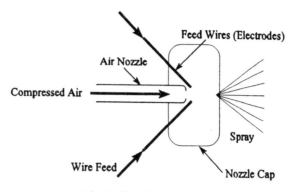

Fig 2 - Electric arc spray gun

With rapid tooling using metal spraying, the process involved in fabricating a sprayed metal injection mould tool does not differ significantly compared with the use of a traditional pattern (see Figure 3). Stereolithography patterns are particularly suitable since the relatively low temperature of the arc-spraying means that the glass transition temperature of the resin, which is approximately 65°C, is not exceeded.

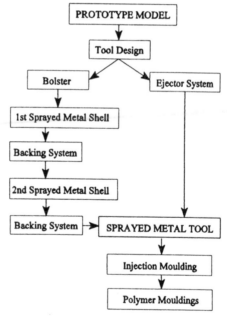

Fig 3 - Sprayed metal tooling process

A range of materials can be sprayed, but by far the most popular materials for mould making are zinc alloys because of their melting point, expansion characteristics, hardness and cost (1).

4.2 Limitations

As well as significant advantages, metal spraying has several limitations when applied to injection mould tooling that require directed research if it is to be used as a competitive technique for producing a wider range of consistent, more durable tooling. As mentioned above, moulding materials requiring high injection pressures or very high moulding temperatures in excess of 250°C can prove problematic and certainly shorten the life of a metal sprayed tool. Although metal spraying replicates fine detail reasonably well, certain intricate features cannot be accommodated. Where small deep features such as holes or recesses exist in a part, these must be filled with metallic inserts, which have to be machined thus increasing the overall cost and lead time for the tool. More importantly, even the use of inserts is not practical, for example, when features such as deep recesses exist in close proximity to each other, the spray is unable to penetrate between the inserts. This is known as a 'shadowing' effect. For example, with two 20mm deep recesses, the minimum distance would need to be at least 10mm.

In addition to physical limitations, commercial awareness and confidence in metal spraying is somewhat poor. This is most probably due to the durability of the tool; high volumes of up to 3 000 components are often quoted in the commercial literature (1) but realistically the numbers produced are considerably lower. Also tools tend to be variable in their performance which is due in the main to the variability in the microstructural properties of the sprayed metal, such as porosity levels, oxide levels and particle size. The main advantages and disadvantages of sprayed metal tooling are summarised in Table 1.

Table 1 - Main advantages and disadvantages of sprayed metal tooling

ADVANTAGE	DISADVANTAGE
COST (Typically 50% of conventional tooling)	LIMITED DURABILITY (harsh moulding conditions degrade the tool)
LEAD TIME (Typically 1 week)	LIMITED GEOMETRICAL COMPLEXITY
REPRODUCTION OF SURFACE DETAIL	LOW ACCEPTANCE IN INDUSTRY . (Lack of confidence and awareness)
CAN ACCOMMODATE COMPLEX SURFACES AND COMPOUND CURVES	INCONSISTENT RESULTS

Research into improvements to the arc-sprayed mould making technique is ongoing especially with reference to the rapid fabrication of prototype, short-run and even production mould tooling (10). The development of automated spraying systems is a priority for a number of research teams (11) to address the variability of spray application; these systems are essentially still test-bed processes. Tool steels have been sprayed but the residual stresses developed within their thickness must be relieved. The 'MUST' process, which uses a combination of spraying and shot peening to simultaneously relieve the stresses, is attempting to address this problem (12). Work has also been done with arc-sprayed steel tooling and an improvement in tool life over arc-sprayed zinc tooling of at least a factor of five was obtained (13). However, there is a surface wear problem with the steel tooling which is being addressed by additional surface treatment.

If arc-spraying is to be used successfully for injection tool making, particularly using zinc alloy, it is essential that the process be fully understood. Many commercial applications have been carried out but too few in depth case studies are available in the literature. For example, the interaction of process parameters with tool performance is not fully understood.

The main process parameters are stated below, together with their influence on the sprayed metal shell. The manufacturers of metal spraying equipment provide recommended settings for most of these parameters based on previous case studies and practical experience. However, an understanding of the effect of each parameter and their combined effect has not been documented and it is felt that research should be directed primarily towards identifying these .

- Arc voltage - the arc voltage influences the amount of heat within each particle and may burn off volatile elements from the wires. Hence the composition of the sprayed shell may be different to the original wires. (19-21 volts is usually recommended for a stable arc)
- Arc current - the current determines the material deposition rate. Increasing the current increases the deposition rate and the heat input to the pattern. (100 amps is typically quoted)
- Wire feed rate - this is linked to the arc current so that if the arc current is changed, the arc spraying equipment automatically adjusts the wire feed rate to suit the selected current.
- Atomising air pressure - this affects the size of the particles being sprayed. Increasing the atomising pressure decreases the particle size and conversely. Particle size may also be influenced by the nozzle and cap configurations.
- Stand-off distance - this is determined by the operator and is defined as the distance between the tip of the spray gun and the pattern. A typical recommended value for stand-off is approximately 200 mm. This is particularly important for avoiding excessive heat input and associated thermal stresses.

There are many more parameters which may affect the properties of the sprayed shell, including the particular material being sprayed, the pattern material, the amount of heat transfer between the sprayed metal and the pattern and also, to some extent, the atmosphere in which the spraying is performed. However, it is only recently that any significant research has been carried out (14) into the effects of all these parameters with reference to the required injection mould tooling properties.

As part of a Brite-Euram II CRAFT project entitled 'Low Cost, Short Run and Prototype Tooling' carried out in conjunction with PERA, the arc-spraying process is being investigated as a rapid injection mould tooling process. The interaction of the spraying parameters with the other tooling components, such as the backing material, is being studied in parallel with the manufacture of a number of injection mould tools. The following section describes the manufacture and moulding of a simple tool. It is hoped that producing tooling with detailed observation will highlight many of the problems and inconsistencies of sprayed metal tools.

5. SPRAYED METAL INJECTION MOULD

The particular tool was a two-impression injection mould for the production of a bearing clamp plate, a sketch of which is shown in Figure 4. The assessment criteria for this tool were :-
- Quality of mouldings - shrinkage and surface reproduction
- Numbers off
- Failure mode(s)
- Thermal performance

The patterns used for the tool were production mouldings in acetal copolymer. Copper heating and cooling pipes were cast into the back of both mould halves and, in order that the mould surface temperature could be monitored during the injection moulding cycle, thermocouples were cast onto the back of the metal shell.

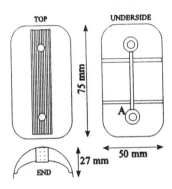

Fig 4 Sketch of bearing clamp plate component

Table 2 is a summary of the manufacturing times and costs for the tool and it is clear that the overall time and cost are significantly lower than for a conventional machined aluminium tool, where the time to manufacture would be at least double and the cost at least three to four times.

The moulding was carried out on a Battenfeld 110 tonne injection moulding machine at PERA. It is generally accepted that clamping forces and moulding pressures should be minimised for sprayed metal tooling to reduce the risk of tool failure and hence a relatively low clamping force was used, initially 10 tonnes, and the material was injected with great care, i.e. the shot weight was gradually increased. The first 100 shots were in Acrylonitrile Butadiene Styrene (ABS) and at this stage no evidence of wear or failure was observed. The ABS was then purged out using a standard purging compound and replaced with acetal copolymer, producing a further 127 mouldings before the metal shell was observed to have cracked on a corner of the mould core, i.e. position A indicated on Figure 4. This gave a total of 227 shots, that is 454 mouldings, before failure. It was considered that failure was due to zero draft on the ribs of the moulding. This design feature is often quoted in the literature as being a major factor in failure of tools but insufficient guidelines are provided. It is apparent from this tool and work that is currently being carried out that generous draft angles of between 3 and 5 degrees will contribute to the success of a tool.

Table 2 Manufacturing times and costs for the sprayed metal tool

Process	Time	Cost
Machine aluminium bolsters	8 hours	Aluminium - £130
Machine melamine faced board	30 mins	Board - £0.50
Machine inserts, ejector pins and sprue bush	4 hours	Metal bar stock - £5
Prepare pattern (side 1)	30 mins	
Spray side 1	40 mins	Zinc alloy - £35
Mix resin and cast backing (side 1)	15 mins	Resin (and filler) - £20
Curing time	45 mins	
Prepare pattern (side 2)	1 hour	
Spray side 2	45 mins	Zinc alloy - £35
Mix resin and cast backing (side 2)	15 mins	Resin (and filler) - £10
Curing time	45 mins	
Demould, clean up and polish	1 hour	
Assemble onto back plates, including machining work for ejector pin holes	1 hour	Standard mould kit - £600
TOTAL	**19 hours, 25 mins**	**£ 835.50**

The shrinkage of the mouldings was measured and it was found that for the ABS mouldings, an overall shrinkage of 0.8% was obtained and for the acetal mouldings 3%. These values are as expected for both materials and hence the sprayed metal tool does not seem to have a significant effect on the material shrinkage.

Table 3 lists the moulding parameters used for both the ABS and the acetal mouldings. It was important to monitor the temperature of the mould surface during moulding as this gives an indication of the effectiveness of the cooling pipes and also the thermal performance of the sprayed metal tool. Cold water was used to cool the cavity when moulding with ABS and the mould surface temperature did not exceed 50°C. When moulding with acetal the cold water cooling was switched off to allow the mould surface temperature to increase, which is usual for this particular polymer. The tool surface temperature increased until it reached a steady state temperature of about 60°C, this is lower than recommended for acetal, 80°C being more typical. This particular tool was not heated because of concerns regarding the effect on the tool materials, particularly the backing, but it became evident that heating the tool would produce better quality mouldings.

Table 3 Moulding conditions

Parameter	ABS	Acetal
*Mould surface temperature, °C	40 - 50	60
Melt temperature, °C	240	200
Clamping force, kN	100	200
Injection speed, mm/s	15	Profiled 30/15/5
Packing pressure, Bar	400	500
Packing time, s	10	10
Cooling time, s	20	20

* Measurement taken from thermocouples cast into the back of the tool.

Some flashing of the mouldings became evident on their long edges when the holding pressure was introduced, however this was not reduced by increasing the clamp force.

With acetal mouldings it was found that the best mouldings in terms of surface finish and reproduction were achieved using a melt temperature of 200°C, a profiled or variable injection speed and a holding pressure of approximately 500 bar. This holding pressure was measured using a transducer at the machine nozzle. There were still some problems with surface finish, notably glossy patches and gate swirl which are characteristic problems with acetal mouldings even using conventional tooling.

6.   CONCLUSIONS

From the case study of this simple tool it is evident that the overall time and cost is significantly lower than for a conventionally machined aluminium tool, where the time to manufacture would be at least double and the cost at least three to four times.

It would seem from the literature and initial tool trials that the behaviour of the sprayed metal shell dominates the durability of the tool. The interface of the shell with other aspects of the tool's construction, such as the bolster and the backing system, will also determine the success of the tool but it is the shell that is of key importance and hence the eventual goal of this research is to be able to produce a sprayed zinc shell with minimal porosity, oxide inclusions and very high surface quality to provide the best possible shell for inclusion in the total sprayed metal tool 'system'.

The next stages of the project are looking in detail at process parameters and the effect of change of these parameters on the final tool. This will be achieved by the fabrication of many different injection moulds utilising the parameters and settings optimised by experiment. Additionally, different surface treatments are being investigated to further improve the durability and consistency of the sprayed metal tools.

It has also become apparent that the moulding conditions selected influence the life of the tool. Thus, it is a necessary part of the research work that guidelines for moulding are developed alongside more durable tooling.

6.   REFERENCES

(1) THORPE, M.L. Progress Report : Sprayed Metal Faced Plastic Tooling. Application Data, File 2.5.1, Issue No. E10203, Tafa Corporation, Concord, USA, 1988.

(2) WOHLERS, T. Future Potential of Rapid Prototyping and Manufacturing Around the World, Keynote Speech, Proceedings of the 3rd European Conference on Rapid Prototyping and Manufacturing, University of Nottingham, UK, July 6-7 1994, pp. 1 - 11.

(3) UPTON, J. TROMANS, G. WIMPENNY, D. STYGER, L. Tooling : The Future of Rapid Prototypes, Proceedings of the 2nd European Conference on Rapid Prototyping and Manufacturing, University of Nottingham, UK, July 15-16 1993, pp. 131 - 141.

(4) SACHS, E. CIMA, M. ALLEN, S. WYLONIS, E. MICHAELS, S. SUN, E. TANG, H. GUO, H. HARTFEL, M. Injection Moulding Tooling by Three Dimensional Printing, Proceedings of the 4th European Conference on Rapid Prototyping and Manufacturing, Belgirate, Italy, June 13-15 1995, pp. 285 - 296.

(5) Central Statistical Office, Business Monitor PAS 3222, 1991.

(6) MUELLER, T. Applications of Stereolithography in Injection Moulding, 2nd International Conference on Rapid Prototyping, Dayton, Ohio, USA, June 23-26 1991, pp. 323 - 329.

(7) BAK, D. Quick Path to Prototype Tooling, Design News, June 25 1990, pp. 116 - 117.

(8) SIMMONDS, R. KAARST-BUTTGEN Electric-arc Metal Spraying With Low-melting Alloys, Kunststoffe (German Plastics), Volume 79, Number 3, March 1989, pp. 5 - 6.

(9) THORPE, M.L. Thermal Spray Industry in Transition, Advanced Materials and Processes, May 1993, pp. 50 - 61.

(10) FUSSELL, P.S. WEISS, L.E. Steel-based Sprayed Metal Tooling, Engineering Design Research Centre, Carnegie-Mellon University, USA, 1990.

(11) WEISS, L.E. LEVENT GURSOZ, E. PRINZ, F.B. FUSSELL, P.S. MAHALINGAM, S. PATRICK, E.P. A Rapid Tooling Manufacturing System Based on Stereolithography and Thermal Spraying, Manufacturing Review, Volume 3, Number 1, March 1990, pp. 40 - 48.

(12) ROCHE, A.D. The Must Process for Producing Prototype Steel Tooling, Proceedings of the 2nd European Conference on Rapid Prototyping and Manufacturing, University of Nottingham, UK, July 15-16 1993, pp. 143 - 156.

(13) WEISS, L.E. THUEL, D.G. SCHULTZ, L. PRINZ, F.B. Arc-sprayed Steel-faced Tooling, Journal of Thermal Spray Technology, Volume 3, Number 3, September 1994, pp. 275 - 281.

(14) FUSSELL, P.S. KIRCHNER, H.O.K. PRINZ, F.B. Sprayed Metal Shells for Tooling Improving the Mechanical Properties, Proceedings of the Solid Freeform Fabrication Symposium, University of Texas at Austin, Texas, USA, August 8-10 1994.

rapid ceramic tooling system for prototype plastic jection mouldings

BETTANY and R C COBB
versity of Nottingham, UK

1   ABSTRACT

The intention of this paper is to describe a newly developed ceramic tooling system, designed specifically to verify new products before production tooling. A major benefit of this process is that the ceramic can be cast directly onto Stereolithography models, providing integration with Rapid Prototyping techniques.

This system can be extremely cost effective, as it does not need expensive equipment, which means that it could be used 'in house'. The work so far has been directed towards producing tooling for injection moulding and case histories are presented to illustrate the success and future potential of this rapid tooling system. The two ceramic systems investigated have demonstrated fast production times, surface reproducibility equivalent to sprayed metal tools and lower costs.

2   INTRODUCTION

Over the past six years, Rapid Prototyping (RP) has been used for a wide range of applications providing compression of 'time to market', parts that can be evaluated for form, fit and function and models for testing.

Recently, one of the major applications of RP has been producing patterns for tooling . The possibility of a production tool failing can be considerably reduced by creating a prototype model and maybe from that a prototype tool. This means that mouldings can be tested thoroughly with as little extra development costs as possible(1). The U.K. tooling market has been estimated at £386m with exports of £62m and imports £146m (2). Typical lead times range from 1 - 26 weeks and here rapid tooling systems offer significant benefits in lowering costs by up to 20% and reducing tooling lead times by 20 - 80%.

Most tooling applications of RP are concerned with plastics injection moulding (3). The main prototype tooling systems that are being used for the injection moulding process at present are given below in descending order of volume of parts.

- Machined Aluminium or Brass, with a lead time of approximately 4 weeks and a capability of producing 50 000 mouldings with good surface quality.

- Electroformed Nickel tools, with a lead time of 4 weeks and a capability of producing 20-500 mouldings with optimum surface qualities with easy removal of plastic parts (4). 5 000 - 10 000 mouldings can also be achieved for the resin transfer moulding process (5).

- Metal sprayed tooling, with a lead time of 1 - 5 days and a capability of producing 50 - 3 000 mouldings with a good surface quality and up to 30 000 mouldings in Low Pressure Styrene (6).

- Resin tools and resin backed photopolymer resin shells, with a lead time of 1 - 8 days, and a capability of producing up to 20-500 mouldings with variable surface quality (3).

There are two distinct categories identified here, the 1 -5 days lead time group which can be categorised as Rapid Tooling. Then there is the 4 weeks lead time group which fits into the low run production tool group. The term Soft Tooling is commonly used to describe all of these systems and any others with the exception of conventional hard steel tooling. It must be noted that all of these tooling systems need a model, with the exception of the machined aluminium which normally would require computer aided design data.

An important point to be aware of is that these soft tooling systems are not as durable as hard steel tooling. It is usual to exaggerate the draft angles and pay particular attention to ribs and sharp radii. Sliding cores are to be avoided, as are high pressures and high temperatures. These are not exactly rules but more like guide lines, to enable as many mouldings as possible to be produced from a tool.

As illustrated in Figure 1, which has been compiled from the author's tool-making experience, the difference in lead times between conventional hard steel tooling and soft tooling systems is certainly evident. It should be noted that the figure also contains values for ceramic tooling, which will be described later.

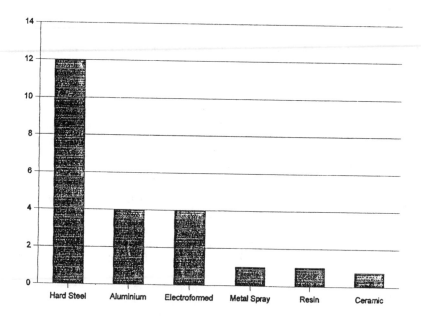

Fig 1 Lead time comparison of tooling types in weeks

3 CHOOSING THE LOWEST COST RP TOOLING SYSTEM

The requirement for prototypes to be made in the same material as the final production part is becoming critical. The emphasis is also on 'rapid' and for very low cost. The fact that most companies produce a limited amount of prototypes over a two year period, usually eliminates any system with a high capital outlay. This leaves resin tooling but also ceramic tooling as the main possibilities for in-house tooling.

Tool making resins usually have a cure time of at least 16 hours, which means that a tool would realistically take 2-3 days to complete. This assumes that the bolster, ejection system and pattern are available. The main drawback of resin tooling is the health and safety aspect, as the installation of extractor and filter equipment can be expensive.

A comparison of tooling costs is illustrated in Figure 2 (these costs have been arrived at from the author's tool-making experience). Here again, ceramic tools appear to offer considerable savings, compared with alternative systems.

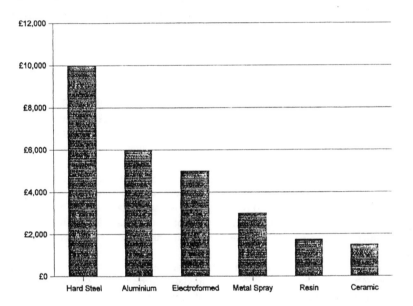

Fig 2 A comparison of typical tooling costs

4 CERAMIC TOOLING SYSTEMS

There are only a few ceramic materials at present on the market, that are specifically aimed at the toolmaking sector of the plastics industry. They are not generally aimed at producing any great quantity of mouldings. However, the relatively short lead times (1 - 5 days) and low tooling costs shown in Figures 1 and 2 make the methods potentially attractive for rapid tooling of parts with volumes of 20 - 500 mouldings although they are known to generate mouldings with variable surface quality.

Comtek is a Chemically Bonded Ceramic (C.B.C.) material produced by the CEMCOM Corp., which has been successful in tooling applications. These C.B.C. tooling materials are Calcium Silicate based and are castable materials that are formed at room temperature. Variable amounts of metal fillers are used to regulate thermal conductivity and fibres are added to enhance the flexural strength (7). Other materials have been developed around Magnesium Aluminium Phosphate binders which have also seen success. Densit is another ceramic tooling system, this system uses micronised Portland cement as a binder.

The objective of this paper is to describe a range of new ceramic materials that have been developed at the University of Nottingham and to compare their properties for toolmaking with commercially available ceramic systems and other rapid tooling systems. The main focus for this work was to reduce the curing time of the ceramic since the commercially available ceramic systems require curing times in excess of 24 hours. In addition the surface quality needs to be improved.

The ceramic system that is described here has a choice of two materials both of which can be cast and demoulded in 2 hours or less. This enables a tool to be made in one day, providing the bolster, ejection system and the model are available. The health and safety aspect is also favourable, as only minimal precautions are necessary.

5   CERAMIC MATERIALS F & J.

For reasons of commercial confidentiality, the two materials described in this work are referred to as ceramic F and ceramic J. They were initially developed to satisfy the market where metal spraying was not suitable. For example, where part geometries possess features such as projections on the model partially blocking the metal spray configuration, or recesses which may be too deep to spray into.

The work was directed towards producing an inexpensive, quick and disposable mould tool, with simplicity of application in mind. Many test pieces were made before arriving at the present formulation. The first trials investigated the possibility of using Magnesium Chlorides, then water extended resin combinations, High Alumina Cements and Gypsum. Many different fillers such as Aluminium powder, wood flour, micronised Zirconium and various fibres were also tested in various proportions.

Ceramic materials F & J can both be cast easily onto stereolithography resin models and providing care is taken during de-moulding the models can be re-used several times. The exothermic reaction only generates a small amount of heat usually in the region of 50°C which does not affect the model. This is not the case with resins of a similar cure time, as they can generate temperatures as high as 130°C during curing.

The material costs of material F and J are around 50% of a metal sprayed tool and the only equipment needed is a simple vacuum chamber and an oven. The materials can be used without being evacuated, although this is not recommended as voids may occur. They are both machinable and can be worked on by hand.

Material J can be demoulded just 30 minutes after pouring and if cured at 40-50° C can reach full compressive strength in hours rather than days.

Material F can be de-moulded in as little as 1 hour after pouring, and can be cured in a similar way to material J.

Ceramic F and Ceramic J are porous materials, which is not desirable when some of the polymers that are used for moulding can possess very adhesive qualities. Therefore various surface treatments have been investigated, some of which were dry film lubricants, semipermanent release agents, Teflon, Polytetrafluoroethylene (PTFE), Silicone and non-Silicone. It is important that the surface treatment should be a process that is quick to apply, so that there is as little delay as possible in running the tool.

5.1 Compressive Strength

The most important physical qualities that were sought after were compressive strength and surface definition. With regard to compressive strength, two widely used tooling resins, Atlas M130, a methyl-methacrylate resin, and EP180, an epoxy resin, were used as a benchmark as shown in Figure 3.

Three 100 mm cubes of each of the materials were cast without any consolidation aids (some voids were evident). Measurements were taken after full cure and the density of each block was calculated. Ceramic F was calculated to 1846 Kg/m3 - 1874 Kg/m3, Ceramic J to 2110 Kg/m3, M130 to 1846 Kg/m3 - 1874 Kg/m3 and EP180 2616 Kg/m3 - 2642Kg/m3.

The four materials showed reasonably consistent values, Ceramic F between 60-80 MPa, Ceramic J 70-74 MPa, tooling resin M130 57.5-60 MPa and tooling resin EP180 97.5-100 MPa.

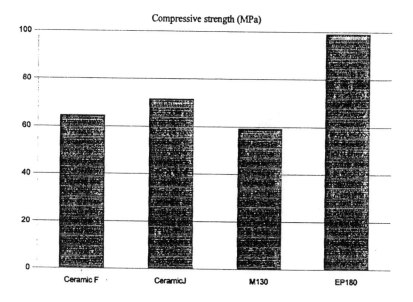

Fig 3 Compressive strength of Ceramic F, Ceramic J and two commonly used tooling resins

5.2 Assessment of Surfaces

Layers of each ceramic were generated by casting onto glass, so that the surfaces could be compared and measured using a Rank Taylor Hobson Form Talysurf. Metal sprayed surfaces from two tool making systems were also generated, again by spraying onto glass as a comparison. The two metal spray systems were Arc-spray and Low-temperature, with zinc alloy and tin/bismuth respectively.

Surface measurements were taken on each sample surface. Each measurement covered a length of 3mm. The mean of five measurements was calculated and then plotted in Figure 4. Ceramic F gave very impressive surface values the mean being 0.128 compared with the tin/bismuth mean of 0.131, which is a common tooling system used because of its surface reproducibility and has an almost mirror finish to the naked eye.

The values for ceramic J were not so impressive as ceramic F or the metal sprayed surfaces, however these values did fall within an acceptable criteria.

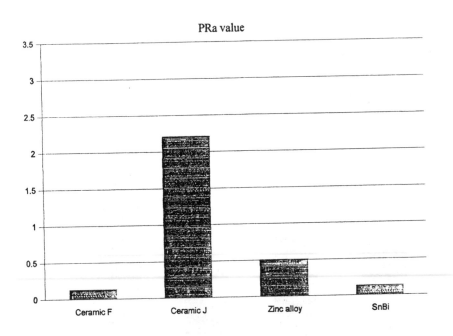

Fig 4 Assessment of surface finish.

6   TOOL J1

Following the encouraging assessment of properties, a tool for producing a perm clip, was manufactured from ceramic J. The tool was designed with overall dimensions of 150mm x 210mm.

The tool J1 was made as one half ceramic J and the other half machined aluminium. Machined aluminium was chosen because the tool was relatively simple on that half and it was as quick to make it that way as it was to cast it in ceramic. Also, it would provide a further tool comparison. A pattern, bolster and backing plates were provided. Figure 5 illustrates the aluminium side of the tool together with a typical moulding.

6.1 The Bolster

The bolster was made from machined aluminium with a brass ejector sleeve in the centre of the moving half. The fixed half had a cavity large enough to facilitate the casting of the ceramic entity. This half also had a plate which bolted on to the back after the ceramic had been ground flat. The sprue was located at the top of the tool. It was 3 mm diameter and 15mm long.

Fig 5 A typical moulding of a Pern clip in position on the moving half of the tool.

6.2 The Process

The powder and liquid components were measured by weight. The powder in this case is added to the liquid unlike mixing cement which is the opposite way round. The mixture could have been either mixed by hand or a slow revolution mechanical mixing machine. It was then placed in a vacuum chamber at 760mm vacuum until the mixture appeared to boil. At this point air was let back into the chamber and it was poured immediately onto the pattern. After 2 hours. the tool was parted and cleaned and the surface was checked for defects. Then it was cured for a further 2 hours. at 40° C. The surface was then treated with acrylic resin and the tool was then taken to a moulder in Nottingham to be run.

6.3 The Moulding

This tool was designed to run fully automatically on an Austin Allen 12.5 ton moulding machine. The machine settings were similar to those used when metal sprayed tooling was to be run. The cycle time was 43 seconds on auto eject, the polymer temperature was 200° C. the polymer being Polypropylene Homopolymer and the release agent being a Fluoroethylene. The machine had a 150mm cylinder a 16mm plunger and the air feed was 80 lbf/in2. which gave a point of injection pressure of 20 000 lbf/in2.

Extra release agent had been applied as a precautionary measure, and the tool ran well producing 400 mouldings before it was taken off the machine. The surface definition of the mouldings was observed to be of reasonable quality. With economy in mind the tool cavity was then knocked out of the bolster, so that it could be used to make a second tool.

To give an idea of time and cost an evaluation was made. The break down was as follows -

- The bolster cost around £130.00.

- The ceramic material £2.00

- Labour cost £110.00.

- The preparation time and casting time came to 4 hours.

  The overall costs, in particular the low material cost, together with the total working time from receiving the pattern compare very favourably with metal sprayed tooling. However, the cycle time for each moulding was relatively long.

7   TOOL J2

This second tool was made in an almost identical way to the first. The difference was that aluminium powder was added to the ceramic in order to increase the thermal conductivity of the material. The increase in thermal conductivity was expected to enable the tool to be cycled at a faster rate, because cooling of the tool would be improved. The surface treatment was also changed from a resin to a PTFE coating to help part release. This PTFE coating has to be baked on at 150°C for one hour. Speed is important in prototyping systems and this cast tool insert took less than a day to complete, including curing of the PTFE treatment.

This second tool was able to be cycled at 30 seconds per shot, showing that the addition of the aluminium filler did in fact improve the thermal conductivity. This tool, which was run by the same moulder, produced 10 740 mouldings before wear was observed on the tool. Figure 6 shows the tool together with the surface damage following moulding. The PTFE coating had to be re-applied only once during the life cycle of the tool. These mouldings were only small and the cavity was only on one side of the tool, this puts the amount of mouldings achieved into a realistic perspective. The cost of this tool was slightly more than the first tool but still far lower than the equivalent metal sprayed tool. These tools are designed to produce a small amount of prototypes, therefore speeding the cycle time up by 13 seconds would not be a significant advantage.

8   TOOL J3

This tool was treated with high temperature silicone oil and was then run by the same moulder but without the incorporation of aluminium powder, since this is potentially explosive.

The bolster was made from machined aluminium, with both halves machined out to facilitate the ceramic entities. The back of the ceramic tools were ground flat to accommodate the aluminium back plates, which were screwed to the rear of the bolsters. Each half of the tool measured 178mm x 100mm x 20mm. Unfortunately this tool only produced 50 mouldings, before developing cracks.

Almost every aspect of the tool itself was the same as the first tool, except for the different surface treatment. Taking time and cost into account even 50 mouldings is a viable proposition and it must be remembered that the bolster and the model were very inexpensive and both could be re-used to produce more tool inserts.

Fig 6 The ceramic half of tool J2, also showing surface damage after moulding

9   TOOL F1

The final ceramic tool was made in Ceramic F, the better surface finish ceramic. PTFE was not used on this tool because it was felt that it would compromise the surface finish of the cast ceramic. Consequently it was thought that another release agent that was under development should be tried; the constituents being high temperature silicone oil and a hydrocarbon material. These were allowed to soak into the porous surface of the cast tool insert.

This tool was run fully automatically and produced over 2 000 mouldings before being taken off the machine. There was no visible deterioration of the tool at this stage. The first polymer to be run in the tool was Acetal co-polymer and the tool temperature was approximately 85 °C. The tool was brought up to the desired temperature by heating it in an oven, with the tool open to ensure as even a temperature as possible.

Polypropylene Homopolymer was injected for the main part of the mouldings, and this was run at a slightly lower temperature. The cycle time was 35 seconds, injection time 5 seconds, hold time 2 seconds, cool time 26 seconds and the tool was air cooled. The tool produced 9 700 mouldings before any significant damage was observed.

10   CONCLUSION

As stated earlier, the two materials were developed initially as an inexpensive and quick method of tooling. The materials are relatively simple to use, with good surface reproducibility and very low cost capital expenditure and show enormous potential for producing accurate tools from rapid prototype models.

The elemental composition of both materials are fundamentally the same, however it appears that their individual performances differ considerably with each of the surface treatments. Since the tools have been proved to run successfully, continued work on the surface treatments is justifiable. This research is being carried out as part of the work-programme of a Brite-Euram Craft project entitled 'Low Cost, Short Run and Prototype Tooling', the research and development being jointly performed between the University of Nottingham and PERA.

The future work will focus on the following :-

* Matching up of surface treatments to the ceramic substrates, to gain maximum performance and repeatability.

* Understanding the limitations of the system and investigating improved surface definition.

REFERENCES

1   BARALDI, U.   EMMERCHTS, C. 'Low Cost Tooling for Injection Moulding' Proceedings of the Composites Tooling 111. 1994, pp. 27-42.

2   Central Statistical Office, Business Monitor PAS 3222, 1991.

3   LÜCK, T. BAUMANN, F. BARALDI, U. 'Comparison of Downstream Techniques for Functional and Technical Prototypes-Fast Tooling with RP.' Proceedings of the 4th European Conference on Rapid Prototyping and Manufacturing, Belgirate, Italy.1995, pp. 247-260.

4   HENTRICH, R. 'Electroformed Mould Cavities and Mould Shells.' Mould Making Handbook for the Plastics Engineer.' Hansler Publishers Munich, Vienna, New York, 1983, pp. 301- 316.

5   BROWN, D. 'The Advantages of Electroform Nickel Tooling for RTM.' Proceedings of the Composites Tooling 111. 1994, pp. 111-130.

6   THORPE, M.L. 'Progress Report: Sprayed Metal Faced Plastic Tooling.' Sales and Engineering Handbook, Tafa Incorporated, Concord, NH, USA, 1988 File 2.5.1..

7   WISE, S. HERSFIELD, C. JONES, K. MELBOURN, T. MILLER, L. RODGERS, T. 'CBC-Versatile Materials for Composites Manufacturing.' Tooling for Composites '90, Anaheim, California, June 5-6 1990.

apid EDM electrodes: linking rapid prototyping to igh volume production

RYALL
ver Advanced Technology Centre, University of Warwick, UK

Synopsis

The development of techniques for producing electrodes for electrode discharge machining (EDM) directly from rapid prototype models, would provide a method of manufacturing high volume production tools. In this way, rapid EDM electrode techniques provide a direct link between rapid prototyping and high volume production.

Various methods for generating EDM electrodes have been investigated at centres around the world. This paper will describe the current-state-of-the-art and utilising examples from the Warwick Manufacturing Group, will explain the relative merits of each of the techniques.

1.0   Introduction

Competition has forced manufacturers to offer a quality product to ensure success. It is now necessary to offer greater choice within the product range or offer something slightly out of the ordinary, a niche product to remain competitive. Greater choice and comparable quality has led to manufacturers offering model updates at progressively shorter intervals. The market life of these products has therefore been getting gradually shorter.

This can be seen clearly in the automotive industry, where to counter competition from the far east, manufacturers are offering new models on a more regular basis. Manufacturers are now looking at means of reducing the time-to-market in an attempt to become more competitive.

One negative aspect of these changes is that tooling, traditionally used for production runs of many hundreds of thousands of components is now only being used for a few hundreds of thousands or less. With a shorter life cycle, the tooling costs have to be amortised over fewer components which increases the base cost of the product or forces a reduction in the profit margin.

Rapid prototyping systems are capable of producing 3 dimensional models directly from CAD data. The models are used as a design aid to prevent mistakes being made early in the design cycle. The models can then be used as a pattern for the production of soft tooling for component testing. This removes the need for prototype hard tooling which may cost almost as much as production tooling. In certain instances soft tooling may be sufficient for low volume production runs in the order of 25000 components. These technologies offer a solution for the production of prototype volumes but are inappropriate for higher production volumes, particularly high pressure processes, (the manufacturing of high volume injection moulding tools or for forging). These circumstances demand the use of more durable tooling materials, more specifically tool steels.

As a consequence, ways are being devised to reduce traditional tooling costs where the application of soft tooling is inappropriate. One of these routes is the application of rapid prototyping to electrical discharge machining. In this process, a high voltage spark erodes the surface of the work piece in a controlled manner, producing the desired form using an electrode. The process is relatively slow in comparison to traditional machining mainly due to the time requirement for the production of the electrode. Electrodes are usually machined in a softer material typically copper or graphite.

Electrical discharge machining (EDM) is used in the production of complicated die geometries to form cavities which are typically difficult to machine, or where the hard nature of the material does not allow the use of traditional machining techniques. The electrodes are made of softer materials where only the reverse form is required, it is therefore easier to machine the electrode than the cavity required in the steel. The process is relatively slow in terms of electrode production and machining. Advances in EDM and its ability to machine unattended throughout the night have seen a steady increase in its use by toolmakers. This paper is a review of techniques currently under evaluation and describes some of the processes and technologies currently under consideration, with an insight into some of the problems and limitations of their use.

2.0 Electro Discharge Machining (EDM)

EDM is a non traditional machining technology where the tool and work piece never come into contact. (1) The process relies on a potential being applied between a tool and work piece in a dielectric fluid. As the distance between the two electrodes is reduced the potential begins to break-down. The break-down potential depends upon, the distance between the two electrodes, the dielectric and the level of pollution between the two electrodes. The electric field generated between the two points allows negative and positive ions to flow ionising the dielectric. A spark is generated as the path between the two electrodes becomes electrically conductive. A by product of this activity is heat which can rapidly achieve temperatures in the region of 8-12000°C, locally melting both the surfaces of the electrode and workpiece.

A bubble forms in the dielectric around the site of the spark, due to vaporisation of both the electrodes and dielectric. The size of the bubble is proportional to the energy being imparted by the spark as it will continue to expand until the level of energy being imparted by the spark equals that being dissipated to the surroundings. When the potential difference is removed the bubble collapses rapidly ejecting molten material from both the work piece and tool. The molten material resolidifies in to small spheres and is flushed away by the dielectric.

2.1 EDM Electrode Materials And Requirements.

For an EDM electrode to be effective it has to be,

- thermally conductive,
- electrically conductive,
- machineable/formeable,
- no re-entrant features,
- have a high melting point.

No single material meets all of these requirements. Commercial materials fall into two categories metallics and non metallics. Although theoretically any material that is electrically conductive could be used, the most commonly used materials are copper, its alloys and graphite. Traditionally electrodes have been manufactured by machining. This route may be relatively lengthy due to the fine detail that originally identified EDM as the best means of producing a cavity. Polishing of the electrode may be necessary to remove any witness marks from machining prior to die sinking.

The production of a master model in the copy milled route has usually been the bottleneck for the production of an electrode.

3.0 Rapid Prototyping Of EDM Electrodes

The application of rapid prototyping to the production of EDM electrodes relies on the production of a pattern. If no data is available for its production it is unlikely that any of the routes described will reduce the lead-time for the production of an electrode. Most of the routes currently being investigated for the production of electrodes have been used or attempted before.

They have previously been found to be to slow or uneconomical due to the time requirement for the production of a pattern by traditional means unless the same geometry is required on a frequent basis.

3.1 Net Shape Casting

Net shape casting is the generation of electrodes by the direct casting of copper to the required form. The basic process requires the production of a pattern for use in the manufacture of a cavity in some other material. Two routes have been used, these are sand casting and investment casting. The latter is favoured due to its ability to produce castings directly from rapid prototype models and in fine detail. Traditionally the pattern material has been wax which is subsequently coated in a ceramic slurry. While still wet the assembly is held over a container with a fine grade of sand. The sand (stucco) is fluidised in a stream of air and adheres to the ceramic slurry. The assembly is then allowed to dry before being recoated in the ceramic and sand until a shell is formed. The shell is then autoclaved to remove the bulk of the wax before firing to remove any residue. This leaves a hollow ceramic shell of the required form. Metal is then cast into this cavity.

Direct Use - investing LOM, SLA, SLS and other types of model has been possible for some time. The process involves the model being connected to a feeding and gating system. The model and tree are then coated until a hard shell is formed. The shell is then fired to remove the model, leaving a cavity into which metal is poured.

Indirect Use - One of the problems associated with the direct use of models is that only one casting can be produced. The cost associated with producing multiple electrodes precludes its use unless an indirect route is chosen. Simple tooling is produced from the pattern into which wax is injected. Two routes have been successfully used by the Manufacturing Group, silicone and rigid resin tooling. Silicone tooling can only be used for the production of simple electrodes and risks the greatest inaccuracies. If wax is injected at to high a pressure the silicone tool may inflate, however to reduce this risk it can be bolstered. Extracting the wax may also introduce further inaccuracies through distortion. Rigid resin tooling requires that draft is applied to the pattern to aid removal. Inaccuracies may be introduced to the geometry during the production of tooling, but will not occur during moulding as with the silicone route.

Cycle times for these routes are significantly longer than that of traditional investment wax tooling due to the poor thermal conductivity of the materials. The wax will also stick to the surface of the resin if the tool heats up to much while in use, even with the use of release agents. To avoid these problems cooling may be incorporated into the mould.

Work within the Manufacturing Group has shown that casting is a perfectly acceptable route, typical turnaround being two weeks for investment casting. Problems have been encountered with this route, mainly through the inexperience of the investment caster in the handling and investing of rapid prototype models. Success is not always guaranteed even with the use of wax patterns, the quality of the electrode may also be too poor for use if correct casting procedures are not followed. High levels of porosity cause arcing of the electrode during EDM machining. It should also be noted that not all geometries are acceptable for investment casting.

3.2   Electroplating

A negative slate filled epoxy or aluminium master is machined or cast and subsequently copper plated. This route has been used for some time and is proven. One draw back of the method is that there are certain geometric limitations. Fine detail may not be possible with geometrically re-entrant features due to local depletion of the electrolyte during plating [2]. The local placement of additional anodes in such areas can reduce this problem

The use of a rapid prototype model directly should significantly reduce the time taken to produce a master pattern, the master pattern being plated directly to produce the electrode. One area of concern is that the form of the shell produced will faithfully copy that of the master pattern. It is essential therefore that the master pattern has all evidence of the laminated build removed before use. The biggest advantage of the process is that the largest electrodes take the same time to plate as the smallest. The only limitation is the size of tank used to plate an electrode.

One method of producing an electrode is described in a paper by 3D systems [3]. The route described uses a stereolithography model as the master pattern. A reversal in silicone rubber is taken. The reversal is then electroless plated to make the surface conductive for the plating process. The silicone form is then plated to a depth of 1.5mm. The shell is then back filled with an epoxy resin before removal from the silicone former to give the electrode some rigidity.

Work on the direct plating of SLA models has also been attempted in the INSTANTCAM (7) project. For this to be successful the model has to be built as a reversal. Both routes have successfully produced electrodes.

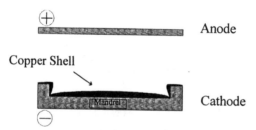

Figure 1. Typical Corner Defect Occurring In A Male Electroform Shell

Electroplating produces reliable accurate electrodes but suffers from a few problems. These are as follows

1) The plating process is non uniform and is generally thinner in deep pockets and corners due to local depletion of the electrolyte, figure 1 illustrates this type of defect (2). These are the same areas in the EDM process which suffer from the highest wear. Burning through in these areas would make further machining impossible.

2) The process is relatively lengthy, typically taking two weeks to plate to a depth of 2-3mm. For small electrodes machining is more viable unless the form is very complicated.

3) Backing of the electrode with an epoxy provides the electrode with the necessary support but creates problems with thermal conductivity in large electrodes. These electrodes are therefore only suited to finishing applications. For roughing operations the heat generated by the process would cause problems with thermal expansion of the shell.

3.3   Abraded-graphite Electrodes

The abrader process works by vibrating and feeding a female cavity in to a block of electrode quality graphite as shown in figure 2. Periodic separation of the tool and work piece is necessary to remove waste graphite powder and to allow the tool to continue cutting.

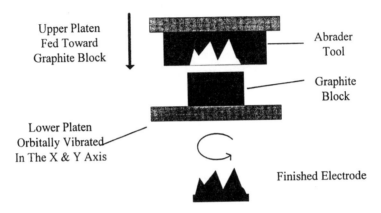

Figure 2: Schematic Representation Of The Abrader Process

Steel tools can be either cut directly by machining the required cavity or by machining an EDM electrode. Filled epoxy tools are cast to the desired shape. These tools have a limited life in comparison to steel tools and are therefore used where low volume production of an electrode is required.

One advantage of this process is that electrodes can be dressed and reused many times before manufacture of a new abrading tool is necessary. The time requirement for the production of an abrader tool could be significantly reduced using an RP model as a master. The master model could be either copy milled or cloned using for example a silicone mould to create an epoxy tool.

This route offers a relatively quick method of producing an electrode but fine detail may well be lost due to the pressures experienced and due to the orbital motion of the process. Allowances have to be made for the motion which would have to be included in the CAD model of the electrode.

3.4   Pressed Copper Electrodes

The pressing of thin copper sheet has only been used where large numbers of electrodes are required as it requires the generation of expensive tooling. The pressing process produces thin walled shells of the desired form in copper. A simple form of pressing where the use of rapid prototypes is appropriate is flexforming.(5) This process only requires the production of a punch, the die being a flexible block of rubber. The punch is forced into the workpiece and forms to the same shape. The rubber provides the necessary resistance to form the workpiece mechanically. There are a few variations on the theme of using a rubber block to press simple forms, a schematic diagram of the process is shown in figure 3. The process has its limitations as the depth of draw is limited to approximately 30-300mm dependant on which method is used and the surface area of the pressing.

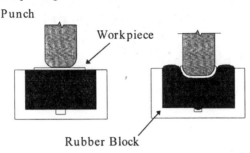

Punch

Workpiece

Rubber Block

Figure 3: Flexible Die Forming

Work has been carried out on a small scale at the technology centre but has shown that the process is relatively straight forward. An epoxy master clone is produced from the RP model and used as the punch. It may be possible to use an RP master directly with some of the newer materials being developed for example the epoxy resins now being introduced for 3D's stereolithography system or a metal pattern from either the Eosint or DTM stations currently under development. Pressing offers a very quick solution to the problem of producing an electrode and is ideally suited to shallow compound curves. For deep features the electrode has to be formed in two or more strikes with an inter-stage annealing operation to prevent work hardening and subsequent shearing of the copper sheet. The material has also to be retained by a clamping ring to prevent creasing of the sheet.

The degree of definition achieved in the work at the technology centre has been surprising with relatively fine detail typically 1mm clearly reproduced in the electrode and used to machine a cavity. This definition is only possible with relatively thin copper sheet (0.5mm), the thicker the sheet used the greater the loss of definition. Thin sheet also has the added problem that it is only suitable for finishing operations. The process has many geometric limitations mainly in the production of sharp/square or very deep features. Metal allowances have also to be considered before production of tooling. The size of electrode produced could be fairly large as the process is used to manufacture low volume/prototype motor body panels. The main limitation is that sharp features which usually identify the need for the EDM process are very difficult to produce by this route. Accuracy of the electrode will depend on the accuracy of the pattern and the degree of spring back of the pressed panel.

3.5 Sintered Electrodes

Keltool Incorporated in the USA manufacture electrodes by a powder metallurgy process.(6) A 1% over size rigid master model is required to produce an epoxy mould. The mould is then filled with a metal powder mix of 68% tungsten and 32% copper and closed under pressure to produce a green compact of the desired form. The electrode is then infiltrated with copper to give a 100% dense electrode.

Keltool claim the electrodes can be produced to within +/- 0.025mm and out perform an equivalent machined graphite electrode in both wear and surface finish. As the electrodes are moulded the process can be used to generate multiple electrodes. The Danish Technical Institute in their work on the rapid production of EDM electrodes state the electrode they had manufactured performed satisfactorily.(7) No details are available on the cost of the electrode but the time requirement for manufacture is one week from receipt of the master model. As Keltool are based in the USA a time penalty would be incurred for both postage of the master model and finished electrode.

3.6   Metal Sprayed Electrodes

Thermal spray equipment has been available for some time in the following forms,

Arc spraying - standard
                      - inert
Plasma-arc spray.
Transferred plasma-arc spray.
High velocity oxygen fuel. (HVOF)

In each process material is melted, atomised and carried at speed on to a substrate, where mechanical bonding of the particles occurs.

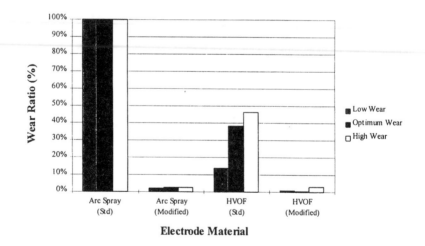

Graph 1: Wear Ratio versus Electrode Material for Different Machine Parameters.

By successive deposition of material a metal shell can be manufactured to the form of the pattern. All of the above systems can be used for the generation of electrodes by spraying copper on to a pattern. Spraying a shell of sufficient thickness, backing of the shell provides a quick solution for the generation of electrodes. The work undertaken in the INSTANTCAM [7] project on the metal spraying of EDM electrodes states that the arc spraying of a pattern is possible. An MCP 137 low melt alloy pattern was produced off an SLA master pattern. The wear performance of the electrodes produced was unacceptably high, being 25% volumetrically. This level of wear or the total failure of the electrodes to machine has been experienced by other workers. Of all the techniques discussed it has the greatest potential for being rapid as an arc sprayed shell can be produced to a depth of a few millimetres in a few hours.

Work within the Advanced Technology Centre on the arc sprayed system initially had no success in producing electrodes. HVOF sprayed electrodes exhibited wear in the order of that seen in the INSTANCAM project. Work has now progressed to a point, with development of the metal spray equipment, where electrodes are now reliably machining steel with volumetric wear of between 1 & 5%. (See graph 1) Further work should see electrodes machining with wear being consistently in the order of 1% or less for no wear EDM machining parameters.

One problem which has yet to be resolved with electrodes from this system is the distortion that occurs due to the contraction of subsequent layers of metal spray. Until this is resolved this particular route will remain closed. Work is being undertaken by Swansea University into the shotpeening of metal spray shells. The shotpeening process impacts ball bearings onto the surface of the metal spray as it is produced relieving the stress. It is difficult to see how this could be applied evenly or effectively in areas of fine detail.

4.0   Conclusions

Rapid prototyping is a relatively new technique which is changing the face of engineering. The techniques described previously are not being used commercially mainly due to problems associated with model accuracy and finish which is reflected in any downstream application. Both surface finish and accuracy is improving but as yet cannot match electrodes produced by machining.

Some of the techniques described have not been tried or require further development. What is clear is there will come a time where they will be actively used. It is unlikely that any one technique will become a clear winner as no one technique can satisfy all the different geometries. All of these techniques may become redundant if the metal sintering techniques under development by EOS and DTM can be used to manufacture electrodes directly in their rapid prototyping machines.

References

1. Electrical Discharge Machining.
J. E. Fuller Rockwell International
Metals Handbook Non Traditional Machining

2. The advantages of Electroformed Metal Moulds In Liquid Composite Moulding

David Brown Ex-press Plastics Ltd
8 Beccles Rd Loddon Norfolk NR14 6JL

3. Manufacture of EDM Electrodes from Stereolithography
3D Systems Technical Publication
3D Systems Inc. Ltd
Unit 7, The Progression Centre, Mark Rd
Hemel Hempstead Herts HP2 7DW

4. ESM Electrode Abrading Machine For The Manufacture Of Graphite Electrodes
Ingersol EDM Innovation Publication, Anon.
Suite 9 Westwood House, Westwood Business Park
Coventry

5. Manufacturing Technology
Lindbeck, J.R. Williams, M.W. Wygant, R.M.

6. Keltool EDM Electrodes
Anon
Keltool Technical Publication
561 Shoreview Park Rd. St Paul, MN 55126

7. Synthesis Report
Author Danish Technological Institute
Brite/Euram
INSTANTCAM
Project No BE-3527-89

The potential for rapid prototyped EDM electrodes using stereolithography models

ARTHUR, R C COBB, and P M DICKENS
University of Nottingham, UK

1   ABSTRACT

The emergence of Rapid Prototyping techniques has speeded up the modelling of components, but its widespread industrial application is limited by the need for a route to production tooling. The changes in the UK tooling market over recent years show a marked increase in the requirement for small tools for moulding of plastics and rubbers. The potential for rapid tooling offers substantial rewards, through reduced lead times and production costs.

A review of research into the application of Rapid Prototyping for manufacture of production tooling is presented. The production of EDM electrodes from Rapid Prototyped SL models is identified as a possible route to mass production tooling and a number of techniques and strategies for an integrated manufacturing approach are discussed. As part of an EPSRC funded research grant an on-going programme of research and development at the University of Nottingham is investigating this area of work, and some initial findings are presented. An appraisal of commercial viability of the technique is also made.

2   A DECADE OF EVOLUTION IN THE UK TOOLING MARKET

Between 1985 and 1992 the UK market for manufacture of small tools increased in value by approximately 32% (1,2,3,4) from £770M to £1018M (Fig 1). Within this industry moulds and dies for plastics manufactured in the UK has been a major growth area, increasing in value by 161% (Fig 2). This substantial increase means that the manufacture of moulds and dies for plastics increased its share of the UK small tool market from 10% in 1985 to 19% in 1992, at £132M (Fig 3), a trend which shows every indication of continuing. This change in market emphasis is in part due to the investment and integration of CAD within even the smallest UK engineering enterprises, providing a platform for competing in the international arena (5).

Electro-Discharge Machining (EDM) is used extensively to produce cavities in hardened tool steel for moulding plastics. It is regarded as an essential facility in todays toolroom, in the form of die sinking and wire cutting apparatus. The EDM cycle can account for 25%-40% of the toolroom lead time in die and mould production, with an even greater percentage of time spent on manufacture of the EDM electrodes for die sinking operations. The diversity of products, reduced life and increasing complexity of components has forced manufacturing times down, as shown in Fig 4 (6).

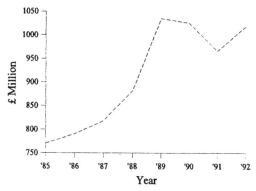

Fig 1 Small tool manufacture in the UK - 1985 to 1992

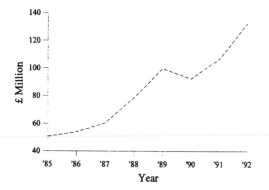

Fig 2 Manufacture of small Moulds and Dies for plastics in the UK

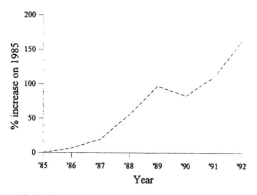

Fig 3 Increasing market for plastics tooling

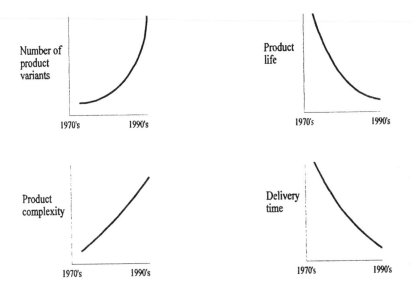

Fig 4 Changes in demands on manufacturing industries
over the last twenty years *(source: Mieritz, B. 1993)*

Recently a range of Rapid Prototyping techniques have emerged to produce highly accurate models which have been used to compress time-to-market, as aids to assess form, fit, and for testing applications. Application to commercial design and development has yielded substantial benefits in reducing lead times and associated costs. Investment in tooling can also be made earlier and with more confidence based upon a proven prototype. This is an extremely important application since typically design and development of a product represents 50% of the overall lead time through to on-line production, the rest of the time being consumed by tooling manufacture. With the increase in die and mould complexity EDM electrode manufacture is becoming ever more time consuming and expensive. There could be considerable savings savings in cost and time by integrating Rapid Prototyping with EDM die and mould manufacture.

3   RAPID PROTOTYPING APPLIED TO TOOLING PRODUCTION

The challenge and greatest potential for Rapid Prototyping lies in providing a direct integrated route to tooling. Methods of 'rapid' tooling manufacture currently used include metal spraying onto models (7) and vacuum casting of silicone rubber moulds (8,9). Both of these approaches provide a quick and relatively cheap route to small batch production and are receiving ever increasing interest as the concept of reduced time-to-market forces industry to look outside the more conventional product development routes. The durability of metal sprayed moulds is presently an obstacle to their wider commercial use. The deterioration of silicone vacuum casting moulds combined with slow cycle times makes large volume production prohibitive. Investment casting of low production quantities within military and aerospace industries is another example of how Rapid Prototyped models have made the transition from concept design to production tools.

Technologists involved in the development of rapid prototyping processes are now devoting much of their efforts to take the philosophy of reduced lead times and development costs through to manufacture of production tooling (9). Between 1991 and 1992 attempts were made to develop applications and techniques for Rapid Prototyped EDM tooling (10,11,12), using StereoLithography (SL) parts. Investigations have also been undertaken with the Fused Deposition Modelling process (FDM) (13) and development work is continuing.

The commercial potential for such a process is huge and it is therefore not surprising that literature is limited, nevertheless details of some elementary experimental work is available (14,15) and although the possible process routes are demonstrated, of more significance are the apparent limitations of the approach which has been adopted. These have concentrated more on the properties of the applied coatings rather than addressing how to optimize the of EDM process parameters.

Generation of a direct production route using SL parts as the electrodes for EDM, achieving substantial metal removal volume combined with low tool wear would have the double impact of unlocking potential of the EDM die sinking process and expanding the role of rapid prototyping in the production environment using SL technology.

4   A CONVENTIONAL APPROACH TO THE PRODUCTION OF EDM ELECTRODES

In the manufacture of dies or moulds using EDM the cost of manufacturing the electrode can represent more than 50% of the total machining cost (16). Any material demonstrating good electrical conductivity can be used as an electrode for EDM, however in practice those with the highest melting point and lowest electrical resistivity are the most efficient.

Traditionally electrodes have been manufactured from metallic materials including various forms of copper, tungsten, brass, and steel, as well as non-metallic material (generally graphites). The conventional methods of producing the electrode profiles include stamping, coining, grinding, extrusion/drawing, and more commonly turning/milling.

The need to reduce electrode manufacturing costs and production lead times is widely recognised and has prompted increasing academic and commercial interest, some of which is specific to Rapid Prototyped models. The possible production techniques may be classified as those which adopt a 'direct' approach to generate a male positive electrode, and an 'indirect' approach requiring a female/negative cavity from which a male/positive electrode is produced (Fig 5).

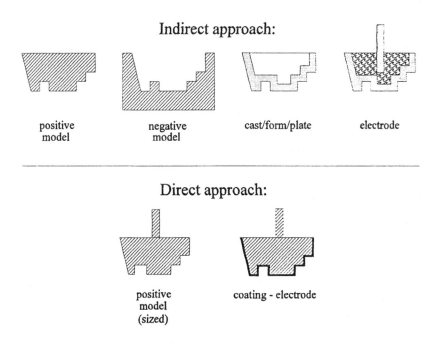

Fig 5 Direct vs. Indirect approach for EDM electrode manufacture
Using StereoLithography models

5   EDM ELECTRODE PRODUCTION USING RAPID PROTOTYPING TECHNIQUES

5.1 Direct electrode production techniques

Three approaches have previously been made to produce electrodes directly from models:-

- Electrically Conductive Plastic (ECP) substrates

ECP 's have been available since 1988 (11) , however, at present they do not have sufficient electrical conductivity. They are not able to sustain the voltages, currents, and high temperatures generated by EDM. At this time it is not viable to use ECP's (12), however further developments in this field may offer some opportunity in the future.

- Metal powder impregnated SL resin substrates

Adding metal powder to liquid resin and attempting to cure the material During the SL process has been evaluated (11,12). Unfortunately it was not possible to cure the composite resin and recognising that dispersal of the powder within the non-conductive resin would not provide sufficient conductivity of the composite model the technique has been dismissed.

- Application of coatings to substrates

Of the many methods available for deposition of material onto a substrate several techniques have been investigated. Metal spraying of copper and nickel onto SL substrates has been explored (14,17,18), and met with limited success however it would appear from the published literature that this technique has not been explored in any great depth, and further experimentation may provide evidence of greater potential. Metallizing SL substrates by electroless plating (14,17) or using Vapour deposition techniques (17) and subsequently electroplating has demonstrated potential and would appear to offer a viable production route. However, reservations have been expressed (19) as to the possibility of achieving a thick enough coating for efficient EDM whilst maintaining model geometries. Various possible routes from SL model through metallizing and coating to EDM electrode have been identified (20).

5.2 Indirect electrode production techniques

Indirect electrode production can be used to generate a thinly coated substrate or a solid electrode, the latter being graphite. The following techniques have been reported:-

- Coated electrodes from a negative pattern

The basic principles of indirect electrode production by using a 'master' model to cast a cavity which is then coated and filled are not new, being established prior to 1975 as a technique for large electrode manufacture (16). These electrodes were used in the automotive industry to finish machine form tools, and the production of large plastic moulds. The positive form is removed and finished, to eliminate imperfections or release agents. A UK patent was filed in 1987 (21) which attempts to cover all configurations of the process for producing electrodes by coating a cavity which has been cast from a model. Various published material describes in some detail the application of Galvanic plating, or Electroforming as it is more widely known, into cavities to manufacture EDM electrodes (11,12,16,19,22). Metal spraying into the pattern cavity as a substitute for electroforming has also demonstrated potential (11,12,16,22), however the porosity of sprayed coatings is blamed for poor efficiency under EDM conditions.

- Solid graphite electrodes from a negative pattern

The two methods currently available for manufacture of solid graphite electrodes via an indirect route are very different. Having produced a negative pattern of the electrode with suitable mechanical properties graphite powder is formed by powder compaction at elevated temperature and pressure (11,12). The 'Hausermann' process performs abrasion of a solid graphite block using vibrating negative patterns (14).

A process known as TARTAN-tooling has also been used for electrode production with promising results (22) however the method of manufacture is not widely used. Another process offering possibilities (22) for which there is little information available is Rotational Copper Casting (RCC) which involves dripping molten copper into a rotating 'negative' cavity.

6   THE WAY FORWARD?

The concept of using models to produce EDM electrodes both directly and indirectly has been researched to differing degrees of detail. There is however no evidence to suggest that attempts have been made to investigate efficiency of the coated electrodes, through manipulation of EDM process parameters. A 3 year EPSRC funded research programme entitled 'Rapid Finishing and Tooling' has recently been set up at the University of Nottingham with the aim of generating EDM electrodes from StereoLithography models. A number of the direct and indirect production methods described above are appropriate to the use of SL models, and require detailed assessment with respect to EDM efficiency. This paper describes some results obtained from electroplated copper layers applied onto silver painted SL parts. A systematic analysis of how EDM parameters influence the efficiency and therefore the viability of rapid low-cost electrode manufacture is also being undertaken.

7   OPTIMIZING EDM DIE SINKING FOR RAPID PROTOTYPED ELECTRODES

The performance of the EDM process is related to the speed of material removal, the sacrificial wear of the electrode, and the surface finish of the eroded cavity. These performance attributes are not complementary requiring the machine operator to select process settings to achieve the necessary balance between them. The general influence of adjustable process parameters is well understood, for a solid electrodes of specific materials. Tabulated data is provided with EDM die sinking machines to guide the user to appropriate settings and the machining characteristics. This information has been generated over the years through, very much, a trial-and-error basis. If thin coated Rapid Prototype electrodes are used the thermal behaviour at the electrode during EDM is likely to be very different from that of solid conductive electrodes. Heat generated at the sparking face is dissipated through the dielectric and into the electrode, but with a thin coated electrode thermal transfer is not as efficient. Delamination of the coatings from SL substrates and rupture of the coating (20) has been found where 'roughing' parameter settings are selected, commensurate with recommendations for solid copper electrodes. What is required to optimize the performance of coated SL electrodes is a detailed examination of the influence of adjustable parameters, from which it should be possible to evaluate the economics and practicality of manufacturing electrodes in this way.

7.1 Process attributes

As mentioned above the performance of the EDM process is measured in respect of machining rate, electrode wear, and surface finish of the eroded cavity. In addition temperature generated at the electrode have been found to significantly influence the behaviour of thin coated SL electrodes. These four attributes which require accurate measurement and optimization are listed in Table 1.

The attributes are conflicting, as optimizing one can have a detrimental effect on the others as demonstrated by the increase in MRR generating higher temperatures and poorer surface finish. The relationship between MRR and TWR is more complex, where both high and low extremes of MRR are detrimental to TWR. A generalised representation of the relationship between the four attributes is described by Fig 6.

Table 1 Measurement of EDM process attributes

Attribute	Descriptor	Measurement	Units
MRR	Material Removal Rate	Volume/Time	mm³/min
TWR	Tool Wear Ratio (electrode:workpiece)	Vol.^{elec}/Vol.^{work}	%
Ra	Surface finish (workpiece)	microns	μm Ra
Temp.	Temperature/Energy at the coating interface	Centigrade	°C

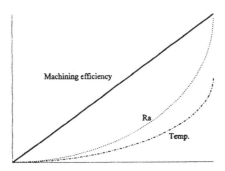

Fig 6a The behavioural characteristics of process attributes
with respect to EDM Machining efficiency

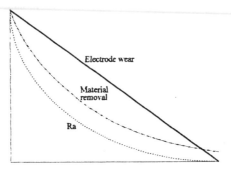

Fig 6b The behavioural characteristics of process attributes
with respect to EDM electrode wear

7.2 Adjustable machining parameters

An the Agie ELOX Mondo 2 die sinking machine is being used for the work, with an oil based dielectric. It has a number of process parameters which can be adjusted by the operator before and during the EDM cycle. Seven parameters have been identified (Table 2) which are to be evaluated in respect of their influence on the EDM process.

Table 2 Adjustable process parameters for optimizing of EDM

Parameter	Descriptor	Measurement	Units
I	Applied Current	Amperes	A
T	Pulse Duration	microseconds	μs
P	Pulse Interval	microseconds	μs
Comp.	Gap Compression	Percentage	%
Gain	Reaction sensitivity of servo.	Percentage	%
Eros. Time	Effective time between electrode 'lift offs'	seconds	s
Rel. With.	Retraction distance for 'lift offs'	millimetres	mm

The Applied Current (I), Pulse Duration (T), and Pulse Interval (P) are generator settings which determine the intensity and pulsing characteristics of the electrical discharge across the spark gap. The Gap Compression (Comp.) and Reaction sensitivity of servo. (Gain) control the servo power on the 'Z', or cutting, axis which resists the increase in pressure within the spark gap caused during machining. During EDM the electrode is periodically retracted from the workpiece, termed 'lift offs', which expels debris from the spark gap area and refreshes the contaminated dielectric. The time between these 'lift offs' and the degree of 'Z' axis retraction are specified by 'Eros. time' and 'Rel. With'.

The extent to which these parameters influence the process attributes is as yet unknown, however their general influence can summarised by Table 3.

Table 3 The influence of process parameters on EDM performance

Parameter	Positive adjustment (beneficial effect)		Negative adjustment (beneficial effect)	
I	High: MRR, TWR		Low: Ra, Temp.	
T	High: MRR		Low: Ra, Temp.	High: TWR
P	High: TWR	Low: Ra, Temp.		High: MRR
Comp.	High: MRR, TWR		Low: Ra, Temp.	
Gain	High: MRR, TWR		Low: Ra, Temp.	
Eros. Time	High: MRR		Low: Ra, Temp.	High: TWR
Rel. With.	High: TWR	Low: Ra, Temp.		High: MRR

7.3 Experimental procedure

• Test methodology using statistical sampling methods

The influence of each of the above parameters and their interrelationships were investigated. To undertake a 'Full Factorial Design' examining all permutations of the seven parameters at three levels would necessitate 2,187 trials, each repeated three times to enable statistical interpretation. It was decided to use Taguchi methods to reduce the number of trials, enabling an initial investigation of parameter influence before proceeding with more detailed examination. Employing a L18 Orthogonal array it is possible to perform a series of 'Fractional Factorial Experiments' (FFE's) which sample the permutations of the seven parameters, reducing the number of trials to 18 (Fig 7). The data from these FFE's should enable identification of the degree of influence and useful range for each of the parameters. Elimination of less critical parameters and limiting the range for the more important ones enables attention to be focused on a detailed programme of tests.

• Preparation of Coated SL electrodes

The SL models used for this experimental series have been designed with a simple geometry, to facilitate surface and volume measurements. Later experiments will employ more complex profiles which are representative of a range of commercial tools. The 'Tablet' models have been built in epoxy SL5170 resin with a layer thickness 0.15mm and designed for easy insertion into an EDM tool fixing (Fig 8). The models were metallized by brush application of a high conductivity silver paint. This silver layer is approximately $10\mu m$ thick exhibiting good adhesion, high electrical conductivity, and a uniform surface distribution. The metallized models were subsequently electroplated with copper, without any further preparation. Initial tests have indicated that a copper thickness of $150-200\mu m$ is sufficient to achieve substantial material removal under semi-roughing conditions. After plating no post processing or finishing was required.

Trial No.	I	T	P	Comp.	Gain	Eros.Time	Rel.With.
A1	4.0	10	6.0	20	15.0	1	1.00
A2	4.0	14	8.5	25	17.5	15	1.25
A3	4.0	18	11.0	30	20.0	30	1.50
A4	7.5	10	6.0	25	17.5	30	1.50
A5	7.5	14	8.5	30	20.0	1	1.00
A6	7.5	18	11.0	20	15.0	15	1.25
A7	11.0	10	8.5	20	20.0	15	1.50
A8	11.0	14	11.0	25	15.0	30	1.00
A9	11.0	18	6.0	30	17.5	1	1.25
A10	4.0	10	11.0	30	17.5	15	1.00
A11	4.0	14	6.0	20	20.0	30	1.25
A12	4.0	18	8.5	25	15.0	1	1.50
A13	7.5	10	8.5	30	15.0	30	1.25
A14	7.5	14	11.0	20	17.5	1	1.50
A15	7.5	18	6.0	25	20.0	15	1.00
A16	11.0	10	11.0	25	20.0	1	1.25
A17	11.0	14	6.0	30	15.0	15	1.50
A18	11.0	18	8.5	20	17.5	30	1.00

Fig 7 The L18 three level orthogonal array used for Fractional Factorial Experiments (23) -
EDM with thin coated StereoLithography models *(source: Taguchi)*

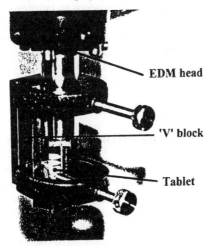

Fig 8 Coated StereoLithography 'Tablet' test piece electrode clamped into a 'V' block
mounted on the EDM head

- Workpiece material

To compare the test results with industrial applications the workpiece material was selected to represent a tooling material commonly used for injection moulding. EN30B through hardened to 51-52HRC was specified for this purpose.

7.4 Experimental results

Selecting three levels for each of the process parameters and applying them to the L18 array, each of the 18 tests were repeated three times, to ensure representative performance. A cut depth of 4mm was attempted for each test. To avoid further variability in the machining conditions no flushing was used. This might be expected to have a detrimental effect on performance, but for these tests the replication of process conditions was of prime importance. The Agie machine is equiped with Adaptive Control but the facility was switched off, again to avoid variability but also to retain control of the conditions at the spark gap with the operator.

For each of the 18 machine set-ups used the coated SL electrodes suffered premature damage and catastrophic failure. However, where a low current was used the surface finish of the eroded cavity was good, i.e. as low as 1.6μm Ra, but machining times were unacceptably high. Under these conditions it would be uneconomic and impractical to remove large material volumes but applications in the re-cutting or finishing of cavities would appear viable. The higher amperage settings, commonly used for semi-roughing and roughing work, achieved greater efficiency through higher MRR but failure of the coatings was a problem. In some cases, particularly with the highest of the three current levels there was substantial damage to the SL model where heat build up had caused melting of the copper coating and distortion of the model.

It would appear from the tests completed so far that heat generation at the front face of the electrode is the primary reason for failure of the coating. Measurement of the thermal conditions generated during EDM is required. The use of thermocouples inserted into the SL model positioned at the interface with the applied coatings has met with only limited success. Electrical interference during the EDM cycle disrupts the signal, but stopping the cut at any given time enables a discrete temperature reading to be obtained. Under semi-roughing conditions the temperature at the interface of coating and substrate has been measured at 62°C. In an attempt to generate 'real time' thermal measurement ungrounded sheathed thermocouple probes have been inserted into SL models, to the interface of coating at the front or sparking face. The problem of electrical interference of the thermal signal still presents a problem.

Analysis of test results is continuing, with an ongoing programme to improve performance of the Rapid Prototyped electrodes.

8   CONCLUSIONS & FURTHER WORK

The initial series of tests described above have provided an insight into the complex interaction of process variables in EDM. It is clear that a greater number of tests are required to identify an optimal machine set up. It is also apparent that the performance and life of the thin coated electrodes is compromised by catastrophic failure, and therefore wear rate (TWR) is not an issue until resilience of the coatings has been significantly improved.

At this point in time it is possible to make limited use of thin coated SL models as electrodes for EDM primarily for re-cutting or finish cutting of cavities, where minimal material removal is required. To make this a viable route to general application it may be necessary to use coatings thicker than 180μm of copper. This may necessitate a scaling or sizing of the SL models to maintain dimensional definition but whether this is practical or economically viable is yet to be determined. Another problem to overcome is that electroplated coatings tend to build up preferentially on external corners and this effect would be more pronounced on increased thicknesses.

The tests are being continued to identify an optimum machine set up with respect to the process attributes listed above. Thicker electroplated coatings are to be applied and tested, with due consideration given to accuracy.

In parallel with this test series other methods of electrode manufacture are being investigated, particularly with respect to indirect production routes using StereoLithography masters. This work will be reported in due course.

9   ACKNOWLEDGEMENTS

The authors wish to thank the following bodies and individuals for their contribution to this paper:- EPSRC, Britains Petite Ltd., Rolls Royce plc (Aerospace Group), J.A.Webster.

10   REFERENCES

(1)   BUSINESS STATISTICS OFFICE. Business Monitor - Quarterly Statistics, PQ3222 Engineers' small tools, First Quarter 1983. HMSO ISBN 0-11-529034-6.

(2)   BUSINESS STATISTICS OFFICE. Business Monitor - Quarterly Statistics, PQ3222 Engineers' small tools, First Quarter 1987. HMSO ISBN 0-11-532322-8.

(3)   BUSINESS STATISTICS OFFICE. Business Monitor, PAS3222 Engineers' small tools, 1991. HMSO ISBN 0-11-535956-7.

(4)   BUSINESS STATISTICS OFFICE. Business Monitor, PAS3222 Engineers' small tools, 1992. HMSO ISBN 0-11-536666-0.

(5)   UK Gauge and Toolmakers Association (personal communication)

(6)   MIERITZ, B. Rapid Prototyping as a Management tool, Proceedings 2nd European Conference on Rapid Prototyping and Manufacturing, University of Nottingham, 15th-16th July 1993. ISBN 0-9519759-1-9 pp 1-15.

(7)   DONNELLY, B. Getting the best out of Metal Spraying, Rapid News. October 1994 pp 10-11.

(8)   DONNELLY, B. Vacuum Casting: The key to Cost-Effective Multiple Prototypes, Rapid News. Spring 1994 pp 6-7.

(9)   LÜCK, T. Comparison of downstream techniques for Functional and Technical Prototypes - Fast tooling with RP, Proceedings 4nd European Conference on Rapid Prototyping and Manufacturing, Belgirate, Italy, 13th-16th July 1995. ISBN 0-9519759-4-3 pp 247-260.

(10)   ALTAN, T. LILLY, B.W. KRUTH, J.P. KÖNIG, W. TÖNSHOFF, H.K. Van LUTTERVELT, C.A. KHAIRY, A.B. Advanced Techniques for Die and Mould Manufacturing, Annals of the CIRP. 1992 Volume 42/2 pp 706-716.

(11)   BJÖRKE, O. (ed.), Layer Manufacturing - A Challenge of the Future, 1992 (Publ. Trondheim). ISBN 82-519-1125-7.

(12)   JENSEN, K. HOVTUN, R. Making Electrodes for EDM with Rapid Prototyping, Proceedings 2nd European Conference on Rapid Prototyping and Manufacturing, University of Nottingham, 15th-16th July 1993. ISBN 0-9519759-1-9 pp 157-165.

(13)   SACHS, E. (et al.) Injection moulding tooling by three dimensional printing, Proceedings 4th European Conference on Rapid Prototyping and Manufacturing, Belgirate, Italy, 13th-16th July 1995. ISBN 0-9519759-4-3 pp 285-296.

(14)   DICKENS, P.M. SMITH, P.J. StereoLithography Tooling, Proceedings 1st European Conference on Rapid Prototyping and Manufacturing,

University of Nottingham, 6th-7th July 1992. ISBN 0-9519759-0-0 pp 309-317.

(15)   MÜLLER, H. EDM Electrodes Made by Rapid Prototyping,
        European Action on Rapid Prototyping. January 1994 BIBA, Bremen, Germany.

(16)   SEMON, G. A practical guide to Electro-Discharge Machining (2nd ed.), 1975
        (Publ. Ateliers des Charmilles, Geneva), Chapter 9 pp 63-76

(17)   SCHULTHESS, A. New Resin Developments for StereoLithography -
        Electroless Plating of SL parts, Rapid Prototyping and Manufacturing conference.
        Dearborn, MI, USA. April 26-28 1994.

(18)   SEGAL, J.I. A study of Metal Sprayed StereoLithographic tooling
        for Electrical Discharge Machining, MSc dissertation December 1994, UMIST.

(19)   CHANTRILL, A. Manufacture of EDM electrodes from stereolithography models,
        1994 (3D Systems Inc. Ltd - UK, Electrode Manufacture 8/94).

(20)   ARTHUR, A. DICKENS, P.M. Rapid Prototyping of EDM electrodes by
        StereoLithography, International Symposium for ElectroMachining (ISEM) XI,
        EPFL Lausanne, Switzerland. April 17th-20th 1995 pp 691-699.

(21)   Patent 2203360A GB. 11 May 1987
        Brian Leonard McKeown, Manufacture of spark erosion electrodes

(22)   INSTANTCAM (BE-3527-89), Deliverable 1&2, Work Area 6 Brite-Euram.

(23)   ROSS, P.J. Taguchi techniques for Quality Engineering, 1988, (McGraw-Hill) pp 228

apid tooling – direct use of SLA moulds for vestment casting

SANG and **G BENNETT**
kinghamshire College of Higher Education, UK

SYNOPSIS

Manufacturing dies for traditional wax investment casting is usually a lengthy and expensive process, especially if the parts are complex. A possible alternative, proposed in this paper, is to cast wax patterns in dies made directly from epoxy photopolymer resin, such as that used for stereolithography (SLA). SLA dies have the advantage of speed associated with rapid prototyping and can significantly undercut the high cost of manufacturing traditional hard tools. The potential application of SLA dies is aimed at low to medium batch production, which effectively bridges the gap between QuickCast^{TM} and hard tooling. A vacuum casting technique was developed to fill the die cavity as opposed to the more common method of high pressure injection. This was done to increase the working life of the die.

1. INTRODUCTION

1.1 RP in Investment Casting

Interest in using rapid prototyping (RP) models in secondary processes is growing steadily as more and more users realise the potential of these techniques. To obtain prototypes in metal, RP models lend themselves readily to investment casting and sand casting with the former being the more popular, since more intricate forms can be cast. Nearly all vendors of rapid prototyping systems have responded to this need by configuring their systems to produce parts suitable for use as investment casting patterns. The obvious advantage here is that no tooling is required, since RP models can directly replace the wax patterns normally used. In investment casting, is the most expensive and time consuming stage of the process is in the production of the hard tooling.

The QuickCast^{TM} build style, developed by 3D Systems Inc., enables stereolithography models to be used directly as sacrificial formers. Characteristically, the enclosed volumes of QuickCast^{TM} models are composed of a honeycomb lattice structure and so have a density which is approximately 20% of that of a solid model of the same part. Models are strong enough to withstand the ceramic shell application process, but the lattice allows the model to collapse away from ceramic shell during thermal expansion. Previous attempts at using fully solid patterns tended to expand and crack the ceramic shell as the former was burnt out. But, even with the use of the QuickCast^{TM} technique, expansion problems are still experienced with larger parts.

DTM's selective laser sintering system (SLS) can build parts using wax and polycarbonate, which are ideal for the investment casting process. Likewise, Stratasys' fused deposition modelling system can deposit a specially formulated investment casting wax (1). Helisys' laminated object manufacturing patterns, although more suited to sand casting, can be burned out of investment casting shells. Other vendors, such as Cubital, Sanders Prototype, and BPM Technologies have built parts for use in certain investment casting processes (1).

An alternative approach is offered by Soligen's direct shell production casting (DSPC) process, whereby investment casting shells are made directly using the layered manufacturing technique. In effect DPSC eliminates yet another stage of the investment casting route, since no tooling *and* no sacrificial patterns are required.

The use of rapid prototyping formers for the investment casting process has made the production of one-off investment castings more affordable and design iterations can be performed without incurring the high cost of re-tooling. However, since RP patterns are considerably more expensive than traditional wax patterns, the cost of making two, three or more patterns of the same part can become prohibitively expensive. Without the economies of scale RP patterns for investment casting do not readily satisfy short or small production runs at reasonable cost.

In comparison, traditional wax patterns are produced by injection moulding with metal mould tools. Such tools are not only expensive to produce but can take many weeks even months to manufacture. This is the main disadvantage of investment casting, since low number production runs will mean a high cost per part. Moreover, the high cost of tooling means that prototyping one casting would not be economically viable for many companies. Committing to tooling is therefore a high risk stage of the production process and many companies hold back until they are satisfied that a design is correct. Changes to the design can mean highly expensive tool re-working once tooling is produced.

1.2 Rapid Tooling

It would clearly be advantageous to have a low cost rapid method of producing small batches of investment cast parts. The logical solution is the compromise combining the low cost rapid aspects of RP techniques with the high volume capacity of traditional methods. That is, use rapid prototyping technology to make the tool or die which can then be used to produce inexpensive wax patterns. One such method involves using a RP master to create a silicon rubber mould into which wax patterns are gravity-cast or injected under low pressure (1). Another is spray metal tooling, which involves spraying molten metal around a RP master to form the die (2). This a longer process but the tools are more durable and can withstand higher injection pressures.

Epoxy photopolymer resin dies made directly from a stereolithography system (i.e. 3D System's SLA 190/20) is the method proposed in this paper. The cost is low and the lead time is short - typically less than one working week including finishing. This rapid tooling method is largely experimental, but the potential is obvious.

2. EPOXY RESIN DIES FOR CASTING INVESTMENT CASTING WAX

2.1 Creation of the SLA die

The objective of the experiment was to design and build a relatively simple, two part, thin-walled epoxy resin mould tool (or die), bolstered in aluminium block. Instead of gravity-casting or injecting the wax patterns, the die cavity was evacuated to remove the problem of air bubbles. Molten wax was then fed into the die under atmospheric pressure.

A switch housing for the Jaguar fighter aircraft was provided by BAe Military Aircraft, Warton as the part for the die. The part was chosen because it was small and only a two part, 'open & shut' die was required. The initial step was to create a three dimensional solid CAD model of the switch, as shown in Fig. 1, from two dimensional drawings. The CAD model was then used as the basis for creating the mould cavity and core of the two mould plates of the die, as shown in Fig. 2 and 3. A shrinkage compensation of 1.6% was added for the wax

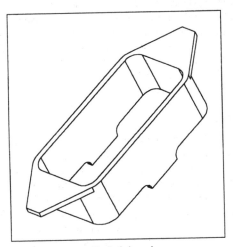

Figure 1 Jaguar Switch housing

based on normal calculations for metal dies. No shrinkage compensation was given for the epoxy dies as this is compensated for in the build parameters. The 3D solid model was created on Parametric Technology Corporation's Pro/Engineer 14 and the mould tool was created using the Pro/MOLDESIGN module. The original mould plates were trimmed so that their overall wall thickness was approximately 2 mm. There were two reasons for doing this: the thin wall reduced the build time and the uniform wall thickness would promote uniform heat transfer and hence a more uniform cooling of the wax.

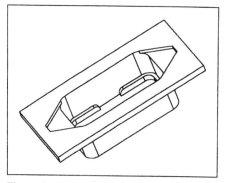

Figure 2 Cavity plate of switch die

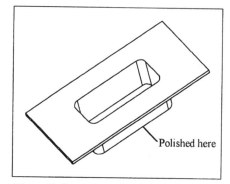

Figure 3 Core plate of switch die

Once completed, STL files of the mould plates were generated. The mould plates were built on a 3D Systems SLA 190/20 at the Centre for Rapid Design & Manufacture (CRDM) at the Buckinghamshire College, High Wycombe. The build orientation was such that the mating faces of the two halve were up-facing as this provided the glass like surface finish required for an airtight seal. The build time for one set of mould plates was approximately 10 hours. Post-finishing, such as support removal, sanding and polishing took approximately the same time as the build. The interior side walls were sanded and polished to ensure that the removal of the wax pattern was not hindered by the stair-stepping effect inherent in SLA models. The stair-stepping on the cavity walls were not sanded away because the wax pattern was expected to shrink away from the walls as it cooled.

2.2 Aluminium bolsters

The next stage was to make aluminium bolster for the mould halves. A diagrammatic representation of the tool is shown in Fig. 4. A block of aluminium with a milled slot was used to support the cavity plate. A channel was also milled around the perimeter of the cavity plate for an 'O' ring seal; this was to ensure that the die was airtight when closed. The protruding section of the cavity was inserted into slot with about two millimetres of overall clearance. This space was for the metal loaded epoxy putty used to bond the cavity plate to the bolster. It was hoped that the metal filler would assist heat dissipation as the wax pattern cooled. Finally, four clamping studs were added to the corners of the bolster to secure the two plates together during wax casting.

For the core plate, a 5 mm thick aluminium block was simply bonded to the top face of the core plate with the epoxy putty (which takes 24 hours to fully set). With the mould pieces in place, four through holes were drilled at each corner of the die for the clamping studs. Two further holes were drilled through the core plate: one for wax entry and one for evacuating the air from the cavity. The construction of the die is shown schematically in Fig. 4 and in actual form in Fig. 5.

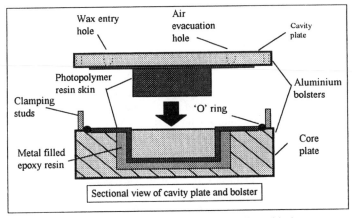

Figure 4 Sectional schematic stereolithography die and bolsters

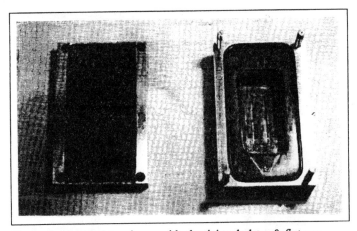

Figure 5 Mould plates shown with aluminium bolster & fixtures

2.3 Wax delivery methods

The most common method for filling metal die cavities with investment casting wax is by high pressure injection (approximately 7 bar). Such injection pressure are necessary to ensure that wax will fill all of the cavity, driving out the air, before it starts to solidify. But in order to withstand such injection pressures the dies have to be strong and, consequently, bulky. The epoxy resin used in making the switch die does not have the same properties as aluminium or tool steel. Being less tough and more brittle meant that it was unlikely to survive the usual injection pressures (later experiments with plastic injection techniques confirmed this). With this in mind, two different methods were considered for administering the wax. One method was injecting the wax under low pressures using a manual injection moulding machine. The other method involved evacuating the die cavity with a vacuum pump so that the molten wax was forced in under atmospheric pressure. As described below, the latter was chosen, since a low pressure injection machine was not readily available.

2.4 Vacuum casting apparatus

A test rig constructed for the 'vacuum casting' method is detailed schematically in Fig. 6. The vacuum pump is not connected directly to the die tool but rather to a ballast tank. The vacuum produced in the tank by the vacuum pump is sufficient to create a (partial) vacuum in the die cavity in only a few seconds. A vacuum valve is used to isolate the die from the ballast tank. This is closed when a sufficient vacuum of approximately 1mm of mercury has been created in the cavity. It also prevents wax escaping out of the die from contaminating the ballast tank and vacuum pump. During the testing phase, investment wax was heated to melting point using an electric heater and a plastic funnel connected to a short (80mm approx.) plastic tube was used to hold the molten wax just before it was fed into the die. A G-clamp was used to close-off and seal the plastic tube when the die cavity was being evacuated. As soon as sufficient wax had been poured into the plastic funnel, the clamp would be loosened and the molten wax would then flow rapidly into the die under atmospheric pressure.

More than twice the volume of wax than was required to fill the cavity was poured to prevent air entering the cavity; the excess wax in the funnel hardened and effectively sealed the die. If air was allowed to infiltrate the cavity during filling then air pockets were formed in the wax resulting in defects and weakening of the wax pattern.

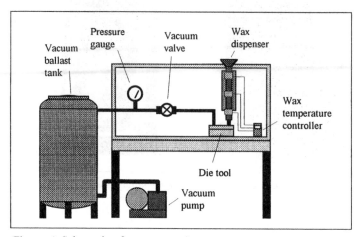

Figure 6 Schematic of vacuum casting apparatus

Initially, this method of dispensing wax was considered adequate for experimental purposes because different investment casting wax formulations could be tested easily and quickly with short changeover times. Also, the set-up allowed for almost immediate testing of the die tool while a more appropriate wax dispenser was being designed and constructed (shown schematically in Fig. 5). In order that the dispenser can be connected quickly to the die an 'O' ring sealed connector sleeve was designed and constructed. The wax is contained in a copper pipe (600mm long, ∅35mm and 1.5mm thick) and is heated above the wax melting point by three 'ring' heating elements. The temperature of the wax is maintained by a digital temperature controller. An integral stirring mechanism along the length of the copper pipe keeps the wax mixture consistent. When the wax has reached the correct temperature, it is dispensed by opening a needle type valve in the 'O' ring connector. As the valve is opened the atmospheric pressure forces the wax into the evacuated die.

2.5 Types of investment casting wax used

Three types of Dussek Campbell investment casting waxes and one speciality modelling wax were tested to determine which formulation was the most suitable for the epoxy photopolymer resin die and for the delivery method used. The Dussek Campbell formulations tested were: Castylene B405 (unfilled), Castylene B489 (filled) and type 289 (filled). The two filled waxes were suited to making production patterns whereas the unfilled wax (B405) was more suited to the production of runner bars and sprues. It was found, however, that this wax produced the best patterns of the three. In comparison to the two filled waxes B405 more readily filled the cavity before cooling and setting. Dewaxing or releasing B405 patterns from the die cavity was also much easier. The filled waxes tended to be more brittle and consequently would fracture more readily. In conclusion greater success was achieved with unfilled Castylene B405, with about a 90% success rate.

In comparison, the speciality modelling wax exhibited similar characteristics to the B405. In general, the patterns created after casting tended to be more flexible and less susceptible to brittle fracture. Consequently, dewaxing tended to be easier as well. Example wax patterns of the switch part is shown in Fig. 7. The blue/green pattern (on the left) is made from the speciality wax and the red pattern (on the right) is the B405 wax.

About 30 wax patterns were produced during the course of the investigation. And when the die was visually examined, no signs of significant wear could be found. There were minor striations on the mating surfaces, however, but these were created during the dewaxing procedure. It can be assumed, therefore, although a further study is needed, that the useful life of the SLA die is significantly beyond 30 shots.

Figure 7 Example wax patterns

2.6 Casting a low melting point alloy and injecting polythene

As a point of interest, two other materials were tested in addition to investment casting wax. MCP 69 a low melting point alloy (69 °C), and BASF Novolen polythene. The MCP 69 alloy (composed of: lead, bismuth, tin and cadmium) was heated in a small oven until it was molten. Then, by following the funnel method described above the molten alloy was poured onto the die cavity under atmospheric pressure. The surface finish of the resulting switch pattern exhibited good edge and feature definition and was comparable to the wax patterns. The alloy pattern has potential applications in EDM machining.

The other material tested was a BASF polythene (Novolen). It was injected at a pressure of 15MPa with a manual injection machine, rather than vacuum cast because the pressure and melt temperature of the polythene (150 °C) was too high for the apparatus to control. As expected, the injection conditions was more than the SLA die could withstand and irreparable damage was caused as a result. The epoxy photopolymer skin and epoxy bond suffered warping and rupture owing to excessive heat and pressure loads. There is a possibility, however, the SLA die could have survived lower injection temperatures and pressures.

2.7 Types of release agent

Release agents used in the investment casting process are designed for a metal die/wax interface, but since in this investigation the interface is epoxy/wax it was uncertain how these release agents would perform. Ambersil silicone based release agents for plastic and rubber parts are common types used in investment casting foundries and two such formulations were tested. These were Formulation One, a heavy duty release agent, and Formulation Six, a general purpose release agent.

Initially, the tool was very difficult to dewax, especially with the filled wax, regardless of how much release agent was applied. Normally, only a very thin coat is needed to dewax a pattern form a traditional metal die. Too much release agent was avoided as this tended to interfere with the flow properties of the wax and leave deformities on the wax surface. This was found to happen with the epoxy tool. As more and more release agent was applied, discernible flow lines began to appear on the surface of the wax pattern. Excessive application of release agent rarely resulted in successful dewaxing.

As expected, more successful dewaxing was attained with the heavy duty Formulation One agent. The thinner coats required reduced the occurrence of flow marks. In conclusion, the combination of Formulation One and either the Castylene B405 or speciality wax resulted in the highest success rate of dewaxed patterns.

2.8 Dimensional Accuracy

A brief examination of the dimensional accuracy of the wax pattern revealed that in general the dimensional errors were approximately +0.4%, for an aluminium casting shrinkage factor of 1.1%. But it is evident that the shrinkage compensation of 1.6%, mentioned earlier, should be adjusted to approximately 1% for the SLA die. Indeed, further tests should reveal the specific shrinkage factor for future SLA dies. As yet, no actual aluminium castings from the wax patterns have been made and so no accuracy results of castings can be presented.

3. TIME & COST COMPARISONS

3.1 Time taken to manufacture the SLA die

An approximate breakdown of the time taken to construct the SLA die is shown graphically in the pie chart in Fig. 8 The overall time is 50 hours, which amounts to almost seven working days. As can be seen in Fig. 8, bonding the bolsters and SLA die parts together comprises 50% of the overall time. In fact, only about an hour if required to prepare the bond, but the epoxy adhesive needs 24 hours to set (as recommended). So the overall time can be significantly reduced by using an epoxy with a shorter setting time.

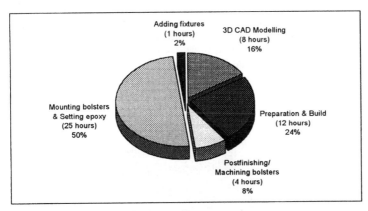

Figure 8 Time breakdown of SLA die construction

On completion of a tool, the next stage is to produce the wax patterns for investment casting. Typically, a lead time of about 2 weeks is required by investment casting foundries. So, for a SLA die, which takes one and half weeks to complete, a prototype casting can be obtained in 3 to 4 weeks. In comparison, traditional toolmakers would need approximately six to 7 weeks to manufacture a hard (metal) tool and produce a casting of the switch housing. With QuickCast™, however, no tooling is required as the pattern is built directly on a stereolithography system. For a foundry with QuickCast™ experience a casting can be produced in 2 weeks. Although nearly all of this time is the lead time required by the foundry since obtaining a SLA pattern takes only 6 hours.

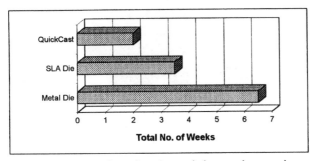

Figure 9 Total number of weeks needed to produce castings

As summarised in Fig. 9, QuickCast™ is the fastest method for obtaining one casting, but, as discussed later, this speed (and cost) advantage ceases to apply as more casting are needed.

3.2 Cost comparison

As well as being the fastest method, QuickCast™ is also the least expensive method of producing a one-off casting as can be shown in Fig. 10. As can be seen, the cost of producing one casting from a QuickCast™ pattern is slightly more than wax patterns (the cost variation is foundry dependent), but this more than offset by the significantly lower cost of producing the pattern initially. Whereas the higher cost of the other two methods is incurred by the need to manufacture dies first before wax patterns can be produced. However, both the time and cost advantage of the QuickCast™ method is eroded as more castings are needed. As shown in Fig. 11, the cost of 20 castings is considerably higher with QuickCast™ because of the high cost of building 20 patterns (about £1400). The production of 20 wax patterns is relatively low since the material costs and mould fill process is very low.

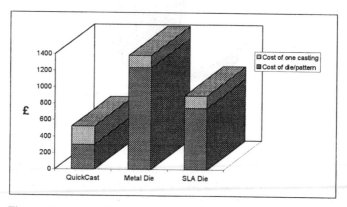

Figure 10 Comparative costs of producing one casting

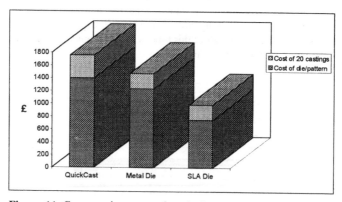

Figure 11 Comparative costs of producing 20 castings

The difference in costs between the SLA die and metal dies are not as distinct. Indeed, the only significant difference is that the SLA die costs about a third less the metal die. Obviously, casting costs are identical, since they follow the same procedure. Cost-wise the SLA die method has a clear advantage but its greatest advantage arises from the shorter lead times for die manufacture.

4. CONCLUSIONS

The task presented in this paper was to investigate the feasibility of directly using SLA epoxy photopolymer resin moulds for small batch production of wax patterns for investment casting. In addition to constructing the SLA die, a vacuum casting system was developed to deliver wax into the die cavity under atmospheric pressure. Successful castings were obtained with the SLA die, but initial dewaxing problems meant that 5 to 10% of the 30 patterns produced were damaged. However, with more user experience and refinement to the tool, a higher success rate should be achievable. Even after 30 shots, no significant tool wear could be found and the tool life was expected to be considerably more than 30 shots.

The SLA die compared favourably against traditional hard tool manufacture both in terms of time and cost. In general, the SLA die took approximately half the time and half the cost to manufacture (but the tool life is not expected to be as long as a hard tool).

The results of the investigation are promising for it demonstrated that for a relatively simple part a die made principally from epoxy photopolymer resin can be used to produce wax patterns. The next stage is to produce dies of greater complexity as this is where SLA dies will have the greatest advantage. Since, unlike traditional tool manufacture where complexity invariably means higher costs and longer lead times, the cost and time needed for SLA dies should not increase significantly.

5. REFERENCES

(1)   Rapid Prototyping Report, 1995,Vol. 5, No. 8, pp3-5
(2)   Jacobs, P.F. QuickCast™ 1.1 & Rapid Tooling, Proceedings of the 4th European Conference on Rapid Prototyping and Manufacturing, 1995, pp1-14, University of Nottingham
(3)   Campbell, J. Castings, 1991, (Butterworth Heinemann)
(4)   Pye, R.G.W. Injection Mould Design, 1983, (George Goodwin)
(5)   Heine, Loper & Rosenthal, Principles of Metal Casting, 1967, (McGraw-Hill)

Physical characteristics of metal sprayed tooling

N DUNLOP
University of Nottingham, UK

SYNOPSIS

Increasing market pressure to reduce costs and lead times has led industry to seek new methods of tooling manufacture. Reduced production runs often mean that conventional manufacturing routes produce tooling which is over-specified and thus unnecessarily expensive. Companies are thus increasingly turning to rapid tooling methods, one of which is the use of metal spraying technologies, for the production of a tooling shell. This methods has been successfully used in the production of tooling for injection moulding, rotational moulding, and pressing operations. The aim of this paper is to describe the physical characteristics of metal sprayed tooling with respect to these production processes, specifically tensile strength, distortion and wear. This paper forms part of an extensive programme at the University of Warwick into all aspects of rapid tooling production.

1.0   INTRODUCTION

With market pressures demanding ever-shorter production runs, with increased product variety, manufacturers are being forced to evaluate the product development process. In the development cycle, tooling forms a major part of the overall product development cost. It is therefore important to develop tooling manufacturing methods which significantly reduce both costs and lead times, when compared to conventional tooling manufacture. One such method is the use of metal spray technologies.

The method of tooling production using metal spray is relatively simple (Figure 1) A master pattern of the desired *component* is produced. This pattern is then mounted onto a suitable base and bolster, and coated with a release agent. A coating of metal spray is then applied to the master pattern, to produce the 'female' form of the desired *tool*. A suitable reinforcement backing is then applied to the shell, depending on the tooling application. Backing materials include Chemically Bonded Ceramics (CBCs), filled epoxy resins, and low melting point alloys. The master pattern is then removed, resulting in the finished tool. In this way, limited production tooling can be produced with substantial cost and lead time savings.

There are a number of metal spraying technologies available [1]; these vary in equipment cost, and quality of coating produced. It is possible to spray a wide variety of materials, from low melting point tin-zinc alloys, to tool steels and nimonic alloys; the quality of the sprayed shell will depend to a great extent upon the system used. Research at the Advanced Technology Centre has centred around two techniques; Electric Arc spray, and High Velocity Oxygen Fuel (HVOF). Electric arc spraying produces economical tooling shells with excellent reproduction and dimensional qualities, but with relatively low mechanical strength and high porosity (<15%). The HVOF process is more expensive, but has the capability of producing coatings with minimal porosity (<1%). The purpose of this paper is to examine the physical properties of these spray technologies, with specific reference to characteristics relevant to tooling. Research to date at the Advanced Technology Centre has been concentrated on the following areas:

- Wear testing, using both benchtop rig testing and injection moulding wear tests. Metal sprayed samples have been benchmarked against other tooling materials.
- Tensile strength of rotational moulding tools. Comparative tests have been performed on samples before and after a number of rotational moulding cycles.
- Distortion testing, to assess the suitability of sprayed materials for tooling applications.
  It should be noted that other areas of tooling applications have been investigated, such as the behaviour of combined sprayed face and backing material under compressive and shear loading. However, this paper is concerned primarily with the behaviour of metal sprayed tooling shells. Subsequent sections will thus go on to describe the research undertaken, and the results obtained. Results have been taken from previous research [2,3] conducted by Msc students in the ATC, and from the current programme of research, which is ongoing.

2.0   WEAR TESTING

2.1 Benchtop wear testing

In this case, a sample of each material was tested under fixed conditions in a test rig (Figure 2). Glass Mat Thermoplastic (GMT) material was selected as a suitably abrasive material for testing, as it is commonly used in moulding applications. In order to simulate the wear occurring during injection moulding, the rig speed was adjusted to provide an average sample surface temperature of 130°C. The wear test was carried out over 5000 cycles - after each 100 cycles, the GMT sheet was renewed, due to erosion of the surface. After each 1000 cycles, the thickness of the metal sprayed sample was measured in order to evaluate different wear rates of materials. Once the test was completed, the sample was measured to give a comparative measure of wear between materials. For benchmarking purposes, typical tooling materials were included in the wear test. The results are shown in Table 1 (the - symbol indicates that a sample did not complete the wear test due to damage).

Sample Material	Wear/1000 cycles (mm)	Overall Rank
Conventional Materials		
Aluminium (HE30TF)	0.33	7
Zinc alloy (Mitsui)	0.01	3
Arc Sprayed Materials		
Tin/Zinc 20/80	-	-
Tin/Zinc 30/70	-	-
Copper	0.26	6
Steel	0	1=
Nickel/Aluminium	0.02	4
MCP Tafa 204M (Zinc alloy)	0.19	5
Aluminium	0.72	8
	-	-
HVOF Materials		
Tool Steel	0	1=

Table 1: Results of benchtop wear testing (taken from [1])

From the results, it can be seen that the wear resistance under benchtop conditions is more related to the hardness of the material itself, rather than the fabrication method used. For example, the two sprayed steel coatings showed negligible wear after 5000 wear cycles, whereas the softer zinc and tin/zinc alloys showed extremely poor wear resistance, with all being destroyed after approximately 200 cycles. The benchtop testing was intended as a general guide as to the overall wear resistance of metal sprayed materials. In order to put the results into a tooling context, the best materials from the benchtop test were used to perform a more representative test, the in-mould wear test, as described below.

2.2 In-Mould Wear Test

As with the benchtop wear test, this was intended as a comparative test. The intention was to insert a number of samples of different materials in a symmetrical mould, and then repeatedly cycle the mould in an injection moulding machine. This would demonstrate the different wear rates on the materials. The mould design is shown in Figure 3. As with the benchtop trials, GMT was selected as the plastic material, with 20% fibre content. This material is extremely abrasive to injection moulds, and thus a useful indicator as to the materials' wear resistance. 1000 injection moulding cycles were performed on the mould. Table 2 shows the results compared with the results from the benchtop wear testing.

Sample Material	Benchtop Wear/1000 cycles (mm)	Ranking	Moulding Wear/1000 cycles (mm)	Ranking
Conventional Materials				
Aluminium (HE30TF)	0.33	7	0.02	3
Zinc Alloy (Mitsui)	0.01	3	0.005	2
Arc Spray Materials				
Steel	0	1=	0.04	5
Nickel/Aluminium	0.02	4	0.08	6
T204M	0.19	6	0.03	4
HVOF Materials				
Tool Steel	0	1=	0	1

Table 2 : Comparison between benchtop and injection moulding wear tests (from [2])

Perhaps surprisingly, the injection moulding results show a considerable change in the overall ranking of the wear rates. Although the rates are generally lower than for the benchtop test, the arc sprayed materials show a marked drop in wear resistance compared to the HVOF and conventional materials. This has been attributed to the higher porosity of the coatings, which allows the molten polymer to penetrate the surface and remove sections of the sprayed surface by an adhesive mechanism during the removal of the moulding. The porous surfaces also have lower resistance to oxidation, thus further weakening the surface integrity in a moulding situation. It can also be seen that HVOF tool steel compares very favourably with the conventional aluminium and zinc alloy materials in the moulding situation. From this, it can be surmised that HVOF tool steel shells are more suited to use in injection moulding situations, when compared to current popular tooling materials, such as T204M. This also would suggest using HVOF coatings in conjunction with softer tooling materials, such as cast aluminium and zinc alloys, in order to provide a more wear resistant surface layer, which could be renewed at will. There are also possibilities for improving the performance of arc sprayed coatings. These include treatment with a PTFE coating, or infiltration with a low melting point alloy, in order to reduce the surface porosity. These methods are currently under investigation. Apart from wear resistance, other trials have been carried out in order to evaluate the effectiveness of metal sprayed coatings - these are described in subsequent sections.

3.0   TENSILE TESTING

The tensile strength of the metal sprayed coating is critical to the success of a tool in certain situations. This is particularly the case for rotational moulding tools, where the metal sprayed shell is not backed with a reinforcing medium. The shell is generally made thicker (<6mm), in order that it may be self-supporting. Softer materials such as 204M (a proprietary alloy by MCP Tafa), Zinc and Tin/Zinc alloys have been used for these tools. These alloys have good dimensional accuracy and surface finish when sprayed, along with very low thermal transfer to the master pattern; thus rapid prototypes may be used as master patterns. Good thermal conductivity is required, to ensure even mould temperatures. The moulds operate at between

250-290°C, close to the softening point for zinc. This may cause increased oxidation of the surface, leading to reduced mechanical strength. From the previous test, it can be seen that the wear resistance of these materials is comparatively poor; however, the working environment is not as demanding as injection moulding. Despite this, it is important that the shell retains mechanical strength after cycling. In order to assess the tensile properties of these alloys, a number of samples were manufactured by spraying onto a flat metal plate, and then cutting the samples to dimensions of approximately 100mm x 20mm x 5mm. Half of the samples were immediately tested to assess the their tensile strength before rotational moulding. The other half of the samples were attached to a production rotational moulding tool, and subjected to 240 moulding cycles. These samples were then tensile tested in the same manner. The results of these tensile tests are shown in Table 3 [3]:

Sample Materials	Cross-section (m^2)	Failure load (kN)	Tensile strength (MN/ m^2)
Before Moulding			
Zinc	107.272.10^{-6}	3.52	32.79
Tin/Zinc 20/80	104.325.10^{-6}	2.78	26.65
204M	107.89.10^{-6}	5.87	54.36
After Moulding			
Zinc	98.86.10^{-6}	2.48	25.04
Tin/Zinc 20/80	102.64.10^{-6}	2.34	22.79
204M	96.67.10^{-6}	3.10	32.76

Table 3: Comparative tensile strength results before & after rotational moulding

As can be seen from Table 3, all three sprayed materials experienced significant degradation in tensile strength, after completing rotational moulding cycling. Despite having the highest initial tensile strength, 204M showed the largest drop in strength, at around 40%. This was probably due to alloying elements in the material; initially giving solid solution hardening, which would dissipate after repeated thermal cycling. In all probability, the reduction in tensile strength was again largely due to the porosity of the materials, allowing increased oxidation at elevated temperatures. As with wear resistance, the best solution would be the application of a surface coating to reduce surface porosity. HVOF stainless steel would be particularly effective in this instance, as it would simultaneously improve the mould wear resistance.

4.0   SHELL DISTORTION MEASUREMENT

This final area of investigation is important, as it determines the dimensional accuracy of the final tooling shell. During the spraying operation, thermal energy from the metal particles will be transferred into the shell material. On cooling, differences in temperature across the surface will cause internal stresses to build up. This will cause distortion of the shell as the spray thickness increases. For this test, a flat copper plate was coated with a PVA release coat. The sprayed material was applied to a thickness of 3mm, with one edge of the sprayed coat clamped to minimise movement. The distortion of the metal sprayed shell was then simply measured using vernier calipers. Each material was tested five times, to ensure repeatability of results. The results of this trial are shown in Table 4.

Material Sample	$T_M(^\circ C)$	Mean Distortion (mm)
Tin/Zinc 20/80	232	0.13
Zinc	410	0.86
204M	422	1.83

Table 4: Results of metal spray distortion test

Predictably, the lowest T_M material showed the least distortion for a given spray depth. The results show that the distortion is not necessarily directly proportional to the T_M. As a rule, however, the higher the T_M of the sprayed material, the higher the thermal energy carried by the metal particles. The stresses built up in the shell will be heavily dependant on a variety of factors, such as humidity, rate of spray, method of component cooling etc. For this reason, these results are merely comparative, and subject to the environmental conditions at the time of spraying. The results are nevertheless useful, as they illustrate the variability of the metal spray process, and the degree of expertise required to successfully spray tooling surfaces.

5.0   CONCLUSIONS

As can be seen from the results, there are a variety of areas where metal spraying can be applied, and a variety of properties which will be affected depending on the tooling application. This paper is by no means a complete research programme for metal sprayed tooling, but is merely intended as an illustration of the technologies' versatility, and the range of properties which need to be considered for tooling applications.   Future work will be concentrated on the application of the HVOF system, for the development of high-integrity tooling shells, with minimum porosity and high mechanical strength - this route will also combine HVOF and Electric arc spray in *hybrid* tooling shells, in order to maximise the benefits of both systems. Surface treatment methods will also be considered, including the use of heat treatment, implantation techniques and plating methods.

6.0 FIGURES

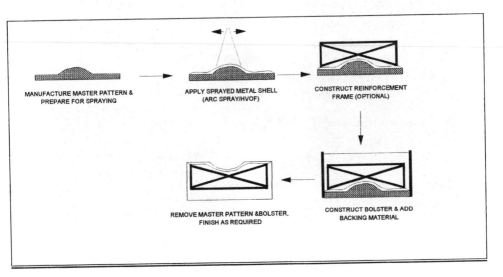

Figure 1: Method of Production for Metal Spray Tooling

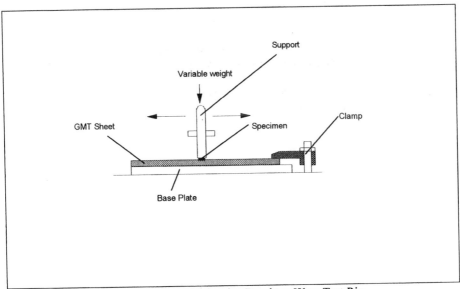

Figure 2: Schematic showing Benchtop Wear Test Rig

Figure 3: Schematic showing mould for injection moulding wear trials

7.0   REFERENCES

[1] Dunlop, RN   'Rapid Tooling using Metal Spray Methods'
                    28th ISATA 1995 Conference Proceedings

[2] Lee, CHC   'Low Cost Tooling for Injection Moulding'
                   MSc dissertation, University of Warwick, 1993

[3] Chang, CS   'Low Cost Tooling for Plastic Forming'
                   MSc dissertation, University of Warwick, 1993